Assisted Eco-Driving

Transportation Human Factors: Aerospace, Aviation, Maritime, Rail, and Road Series

Series Editor: Professor Neville A. Stanton, University of Southampton, UK

Automobile Automation
Distributed Cognition on the Road
Victoria A. Banks and Neville A. Stanton

Eco-Driving
From Strategies to Interfaces
Rich C. McIlroy and Neville A. Stanton

Driver Reactions to Automated Vehicles
A Practical Guide for Design and Evaluation
Alexander Eriksson and Neville A. Stanton

Systems Thinking in Practice
Applications of the Event Analysis of Systemic Teamwork Method
Paul Salmon, Neville A. Stanton, and Guy Walker

Individual Latent Error Detection (I-LED)
Making Systems Safer
Justin R.E. Saward and Neville A. Stanton

Driver Distraction
A Sociotechnical Systems Approach
Kate J. Parnell, Neville A. Stanton, and Katherine L. Plant

Designing Interaction and Interfaces for Automated Vehicles
User-Centred Ecological Design and Testing
Neville Stanton, Kirsten M.A. Revell, and Patrick Langdon

Human-Automation Interaction Design
Developing a Vehicle Automation Assistant
Jediah R. Clark, Neville A. Stanton, and Kirsten Revell

Assisted Eco-Driving
A Practical Guide to the Design and Testing of an Eco-Driving Assistance System (EDAS)
Craig K. Allison, James M. Fleming, Xingda Yan, Roberto Lot, and Neville A. Stanton

For more information about this series, please visit: https://www.crcpress.com/Transportation-Human-Factors/book-series/CRCTRNHUMFACAER

Assisted Eco-Driving

A Practical Guide to the Design and Testing of an Eco-Driving Assistance System (EDAS)

Craig K. Allison, James M. Fleming, Xingda Yan, Roberto Lot, and Neville A. Stanton

CRC Press is an imprint of the
Taylor & Francis Group, an **informa** business

First edition published 2022
by CRC Press
6000 Broken Sound Parkway NW, Suite 300, Boca Raton, FL 33487-2742

and by CRC Press
2 Park Square, Milton Park, Abingdon, Oxon, OX14 4RN

© 2022 Taylor & Francis Group, LLC

CRC Press is an imprint of Taylor & Francis Group, LLC

Reasonable efforts have been made to publish reliable data and information, but the author and publisher cannot assume responsibility for the validity of all materials or the consequences of their use. The authors and publishers have attempted to trace the copyright holders of all material reproduced in this publication and apologize to copyright holders if permission to publish in this form has not been obtained. If any copyright material has not been acknowledged please write and let us know so we may rectify in any future reprint.

Except as permitted under U.S. Copyright Law, no part of this book may be reprinted, reproduced, transmitted, or utilized in any form by any electronic, mechanical, or other means, now known or hereafter invented, including photocopying, microfilming, and recording, or in any information storage or retrieval system, without written permission from the publishers.

For permission to photocopy or use material electronically from this work, access www.copyright.com or contact the Copyright Clearance Center, Inc. (CCC), 222 Rosewood Drive, Danvers, MA 01923, 978-750-8400. For works that are not available on CCC please contact mpkbookspermissions@tandf.co.uk

Trademark notice: Product or corporate names may be trademarks or registered trademarks and are used only for identification and explanation without intent to infringe.

Library of Congress Cataloging-in-Publication Data

Names: Allison, Craig K., author.
Title: Assisted eco-driving: a practical guide to the design and testing
   of an eco-driving assistance system (EDAS) / Craig K. Allison, James M.
   Fleming, Xingda Yan, Roberto Lot, and Neville A. Stanton.
Description: First edition. | Boca Raton, FL: CRC Press, 2022. | Includes
   bibliographical references and indexes.
Identifiers: LCCN 2021026421 | ISBN 9780367532628 (hbk) |
   ISBN 9780367532635 (pbk) | ISBN 9781003081173 (ebk)
Subjects: LCSH: Driver assistance systems. | Automobile driving—Human
   factors. | Automobiles—Automatic control. | Automobiles—Fuel
   consumption.
Classification: LCC TL152.8 .A45 2022 | DDC 629.2—dc23
LC record available at https://lccn.loc.gov/2021026421

ISBN: 978-0-367-53262-8 (hbk)
ISBN: 978-0-367-53263-5 (pbk)
ISBN: 978-1-003-08117-3 (ebk)

DOI: 10.1201/9781003081173

Typeset in Times LT Std
by KnowledgeWorks Global Ltd.

Contents

List of Figures xi
List of Tables xv
List of Common Symbols xvii
Preface xix
About the Authors xxi
Acknowledgements xxiii
List of Abbreviations xxv

Chapter 1 Eco-Driving: Reducing Emissions from Everyday
Driving Behaviours 1

    Introduction 1
    Transportation Contribution to GHG 1
    Eco-Driving 2
    Eco-Driving Knowledge 3
    Eco-Driving Training 6
    Feedback and Eco-Driving 7
    Eco-Driving Feedback in Other Sensory Modalities 11
    Conclusions 14
    References 14

Chapter 2 Applying Cognitive Work Analysis to Understand
Fuel-Efficient Driving 19

    Introduction 19
    Cognitive Work Analysis 20
    Method 22
    Results and Discussion 23
        Work Domain Analysis 23
        Control Task Analysis 30
        Strategies Analysis 33
        Social Organisation and Cooperation Analysis 35
        Worker Competency Analysis 37
    Generating Specifications 39
    Conclusions 44
    References 45

Chapter 3 Adaptive Driver Modelling in Eco-Driving Assistance Systems 49

    Introduction 49
    ADAS for Safety and Eco-Driving 50
    Models of Driver Behaviour 51

v

	Car-Following Behaviour	51
	Cornering Behaviour	53
	Methods	55
	Hardware	55
	Naturalistic Data Collection	55
	Results	56
	Car-Following	56
	Cornering	59
	Discussion	63
	Comparison of Models with Naturalistic Data	63
	Parameters Characterising Driver Behaviour	64
	Limitations	65
	Conclusions	66
	References	66
Chapter 4	Taming Design with Intent Using Cognitive Work Analysis	71
	Introduction	71
	Designing Interfaces	72
	Cognitive Work Analysis	74
	Research Goal	76
	Method	76
	Participants	76
	Procedure	76
	Results and Discussion	79
	Workshop 1 - Waiting at Traffic Lights	79
	Design of the Display	79
	Validation of the Display	82
	Workshop 2 - Accelerating to Overtake	84
	Design of the Display	84
	Validation of the Display	87
	General Discussion	88
	Conclusions	89
	References	90
Chapter 5	Applying Design with Intent to Support Creativity in Developing Vehicle Fuel Efficiency Interfaces	93
	Introduction	93
	Vehicle Fuel Efficiency	93
	Design with Intent Toolkit	95
	Rationale Summary	98
	Case Study	98
	Participants	98
	Procedure	98
	Review of the Ideas and Final Coding	99
	Design Results & Discussion	100

Contents · vii

          Driver Acceptance 103
          Validation Methodology 103
              Participants 103
              Measures 105
              Procedure 105
          Validation Results & Discussion 105
          General Discussion 109
              Use of the DwI Toolkit 110
              Realising Eco-Driving through Design 111
          Conclusion 111
          References 112

Chapter 6   Incorporating Driver Preferences into Eco-Driving
                 Optimal Controllers 117

          Introduction 117
          Literature Review 119
              Optimal Control 119
              Models of Vehicle Following 120
              Models of Cornering Speed 122
          The Driver Satisfaction Model 123
              Development of Cost Function and Constraints 123
              Choice of Weighting Parameters 124
              Incorporation of Cornering Constraints 127
              Comparison of the Model with IDM 129
          Model Validation 131
              Method 131
              Results and Analysis 134
          Usage Example 135
          Conclusion 137
          References 138

Chapter 7   Receding Horizon Eco-Driving Assistance Systems
                 for Electric Vehicles 143

          Introduction 143
          Speed Advisory Problem 145
          Vehicle Motion and Driver Modelling 146
              Vehicle Motion Dynamics 146
              Driver Preference Model 146
          Electric Powertrain Energy Consumption Model 147
              Driving Losses 148
              Powertrain Losses 151
          Full-Horizon Optimisation 152
              Boundary Conditions 152
              Car-Following Case 153
              Cornering Case 155

	Receding Horizon Control 157
	Boundary Conditions... 157
	Lead Vehicle Trajectory Prediction.................................... 157
	Receding Horizon Cost Function 158
	Terminal Cost Selection 159
	MPC without Terminal Cost... 159
	MPC with Terminal Cost... 160
	Test Case Under Real-World Driving Data 163
	References 165
Chapter 8	**In Simulator Assessment of a Feedforward Visual Interface to Reduce Fuel Use** 169
	Introduction 169
	Hypotheses 171
	Method 171
	Design.. 171
	Participants 171
	Equipment and Driving Scenario....................................... 171
	Procedure.. 174
	Measures and Analysis.. 175
	Results 177
	ANOVAs.. 177
	Fuel Usage 178
	Average Speed 178
	Speed RMS Deviation 180
	Mean Acceleration... 181
	Acceleration Time 182
	Mean Braking Deceleration 182
	Braking Time.. 183
	Workload 184
	Discussion.. 185
	Effects on Driving Style 186
	Effects on Fuel Consumption 187
	Effects on Cognitive Workload 187
	Limitations.. 188
	Opportunities for Future Work... 188
	Conclusion 188
	References 189
Chapter 9	**Assisted versus Unassisted Eco-Driving for Electrified Powertrains**... 193
	Introduction 193
	Powertrain Models 194
	Conventional Powertrain 195

	Parallel Hybrid Powertrain	196
	Battery Electric Powertrain	197
	Study Method	197
	Equipment	197
	Study Design	198
	Statistical Analysis	198
	Results	199
	Discussion	201
	Limitations	202
	Conclusions	202
	References	203

Chapter 10 Predictive Eco-Driving Assistance on the Road 205

Introduction 205
System Architecture 207
Perception Layer 208
    GPS-Based Localisation 208
    Long-Range Radar Sensing 209
    Vehicle ECU 209
Decision Layer 210
    Fuel Consumption Model 210
    Driver Preference Model 211
    Predictive Optimisation of Vehicle Speed 213
Action Layer 214
    Visual Interface 214
Simulator Testing 215
    Test Procedure 215
    Results 217
On-Road Testing 218
    Test Procedure 218
    Results 220
Discussion 222
    Limitations 223
Conclusions 224
References 224

Chapter 11 Designing for Eco-Driving: Guidelines for a More
Fuel-Efficient Vehicle and Driver 227

Introduction 227
Chapter Summaries 227
Future Work 230
Summary of Guidelines, by Chapter 231
    Chapter 1 231
    Chapter 2 231

Chapter 3 and Chapter 6 232
Chapter 4 and Chapter 5 232
Chapter 7 233
Chapter 8 233
Chapter 9 233
Chapter 10 233

Author Index 235

Subject Index 239

List of Figures

Figure 2.1	Excerpt from the completed Abstraction Hierarchy.	30
Figure 2.2	Excerpt from the completed CAT analysis.	32
Figure 2.3.1	Simplified Flow map for the task of deceleration to a lower speed.	33
Figure 2.3.2	Simplified Flow map for the task of acceleration to a higher speed.	34
Figure 2.3.3	Simplified Flow map for the task of maintaining headway.	34
Figure 2.3.4	Simplified Flow map for the task of maintaining current speed.	34
Figure 2.4	An excerpt from the completed SOCA-CAT analysis of the agents involved in fuel-efficient driving, including legend.	36
Figure 3.1	Example of video data from ADAM.	55
Figure 3.2	A typical car-following spiral.	56
Figure 3.3	Observed vehicle spacings for F1.	57
Figure 3.4	Observed relative velocities for F1.	58
Figure 3.5	Observed inverse TTC for F1.	59
Figure 3.6	Observed lateral accelerations in cornering data.	60
Figure 3.7	Detail of high-speed data.	61
Figure 3.8	Observed velocities for cornering data.	61
Figure 3.9	Alternative lateral acceleration models.	62
Figure 4.1	Designed HDD interface mock-up for the scenario "Waiting at traffic lights".	79
Figure 4.2	Subset of the Abstraction Hierarchy accounted for by the fuel efficiency display (Item 4, Table 4.3).	83
Figure 4.3	Designed HUD interface mock-up for the scenario "Accelerating to overtake". Within the current image the vehicle has overtaken the vehicle in the middle lane and is being informed they should be prepared to follow the ghost car in moving back to the middle lane.	84

Figure 4.4	Designed HDD interface mock-up for the scenario "Accelerating to overtake". Within the HDD interface the driver, presented in the rear green car is being informed that in 14 seconds it will be safe to overtake the preceding yellow car, and is informed that the optimal overtaking speed is 69 mph in 5th gear. 85
Figure 4.5	Subset of the Abstraction Hierarchy accounted for by the "Accelerating to Overtake" display. 87
Figure 5.1	Number of ideas generated for each of the DwI Lens. 100
Figure 5.2	Number of ideas generated for each feedback modality during the workshop. 102
Figure 5.3	Box and Whisker plot for participants' likelihood to use the proposed interfaces. 107
Figure 5.4	Box and Whisker plot for participants' perceived effectiveness of the proposed interfaces. 109
Figure 6.1	Car following, comparison with IDM for different 'a'. 129
Figure 6.2	Braking, comparison with IDM for different 'T'. 130
Figure 6.3	Example of start/stop test case (Test S4). 132
Figure 6.4	Example of vehicle-following test case (Test F4). 132
Figure 6.5	Example of cornering test case (Test C4). 133
Figure 6.6	Eco-driving optimal control example. 138
Figure 7.1	Electric vehicle structure and losses considered. 148
Figure 7.2	Forces relationship. 149
Figure 7.3	Front wheel braking curve. 150
Figure 7.4	Battery current usage map of electric powertrain. 151
Figure 7.5	Full-horizon solutions under different w [car-following case]. 154
Figure 7.6	Pareto curve under different w [car-following case]. 155
Figure 7.7	A case of road curvature. 156
Figure 7.8	Trade-off solutions under different w [cornering case]. 156
Figure 7.9	Receding horizon control solutions under different w [no terminal cost]. 160
Figure 7.10	Speed profile of the remaining journey. 161
Figure 7.11	Receding horizon control solutions under different w [with terminal cost]. 162

List of Figures

Figure 7.12	Trade-off solutions with ADAM data.	163
Figure 8.1	The University of Southampton driving simulator.	172
Figure 8.2	The driving route used in simulation, arrows indicate direction of travel. © 2021 Google.	173
Figure 8.3	The speedometer used in the assisted eco-driving trial to display the recommended speed range. In the control and unassisted eco-driving conditions, the same display was used, but without the speed recommendation band.	174
Figure 8.4	Participants mean fuel usage for the different locations, by condition.	178
Figure 8.5	Participants' female (5a) and male (5b) mean speed.	179
Figure 8.6	Participants' speed RMS deviation for the different locations, by condition.	180
Figure 8.7	Participants' mean acceleration.	181
Figure 8.8	Participants' mean acceleration time in seconds.	182
Figure 8.9	Participants' mean braking deceleration.	183
Figure 8.10	Participants' mean braking time in seconds.	184
Figure 8.11	Participants' workload as measured by NASA-TLX.	185
Figure 9.1	Powertrains considered in the study.	195
Figure 9.2	Efficiency map of the IC engine.	196
Figure 9.3	Efficiency map of the electric traction motor.	197
Figure 9.4	Fuel usage of conventional powertrain.	200
Figure 9.5	Fuel usage of hybrid powertrain.	201
Figure 9.6	Energy usage of electric powertrain.	201
Figure 10.1	System architecture.	208
Figure 10.2	Contour map of fourth-order fuel consumption model.	211
Figure 10.3	Predicted vs. actual fuel consumption, test data.	212
Figure 10.4	The visual HMI, showing recommended speeds to the driver.	215
Figure 10.5	The fixed-base driving simulator used for trials.	216
Figure 10.6	Fuel consumption in simulator testing.	217
Figure 10.7	Average speeds in simulator testing.	218
Figure 10.8	The instrumented vehicle used for testing.	219

Figure 10.9 Following route. Map data: © 2021 Google. 219

Figure 10.10 Cornering route. Map data: © 2021 Google. 220

Figure 10.11 System response in cornering test. 221

Figure 10.12 System response in following test. 221

List of Tables

Table 2.1	Object Related Processes and Physical Objects Identified within the Abstraction Hierarchy	24
Table 2.2	Description of the Physical Objects Layer, and Their Influence on Fuel Usage	26
Table 2.3	SRK Taxonomy	38
Table 2.4	Specification Insights from Each CWA Stage	41
Table 2.5	Example Case Studies regarding the Insights Offered by CWA	43
Table 3.1	Parameters of the IDM	52
Table 3.2	Observed Minimum Car-Following Parameters	58
Table 3.3	Observed Cornering Parameters	62
Table 3.4	Recommended Parameters for an ADAS Driver Model	64
Table 4.1	Lenses and Themes Presented within the DwI Framework	73
Table 4.2	Workshop Summary	78
Table 4.3	Summary of Interface Elements	80
Table 4.4	DwI Cards Used to Inspire Design of the Fuel Efficiency Display within the "Waiting at Traffic Lights" Interface	81
Table 4.5	DwI Cards Used to Inspire Design of the "Accelerating to Overtake" Interface	86
Table 5.1	Summary of the Lenses and Themes Present within the DwI Framework	97
Table 5.2	Ideas Inspired by the Architectural Lens, including the Specific Cards Used	101
Table 5.3	Summary of the 14 Interface Ideas Generated within the Workshop, Designed to Increase Drivers Fuel Efficiency	104
Table 5.4	Participants Median and Standard Deviation for Likelihood of Use for the Different Interface Designs Developed following the DwI Workshop (where 1 = Very Unlikely and 5 = Very Likely)	106
Table 5.5	Participants Median and Standard Deviation for Perceived Effectiveness for the Different Interface Designs Developed following the DwI Workshop (where 1 = Very Low Impact and 5 = Very High Impact)	108

Table 6.1	Model Parameters for the Comparison	129
Table 6.2	Results for Start/Stop Test cases	134
Table 6.3	Results for Vehicle-Following Test cases	134
Table 6.4	Results for Cornering Test cases	135
Table 7.1	Energy Economy Compared between Whole Horizon and Receding Horizon Solutions	162
Table 7.2	Energy Saving Comparison	164
Table 8.1	ANOVA for Condition, Location, and Gender, Main Effects and Interactions	177
Table 8.2	Post hoc Tests, Condition × Location Interaction for Fuel Usage	178
Table 8.3	Post hoc Tests, Condition × Location Interaction for Mean Speed	180
Table 8.4	Post hoc Tests, Condition × Location Interaction for Speed RMS Deviation	181
Table 8.5	Post hoc Tests, Condition × Location Interaction for Mean Acceleration	181
Table 8.6	Post hoc Tests, Condition × Location Interaction for Acceleration Time	182
Table 8.7	Post hoc Tests, Condition × Location Interaction for Braking Deceleration	183
Table 8.8	Post hoc Tests, Condition × Location Interaction for Braking Time	184
Table 9.1	Vehicle Parameters	195
Table 9.2	Descriptive Statistics	199
Table 9.3	ANOVA Results for Test Condition	199
Table 9.4	Post Hoc t-Tests of Fuel/Energy Usage	200
Table 10.1	Polynomial Coefficients of Selected Fuel Model	210
Table 10.2	Comparison of Polynomial Fuel Consumption Models	211
Table 10.3	Parameters for the Driver Model	213
Table 10.4	GPS Coverage during Test	222
Table 10.5	Radar Tracking during Test	222

List of Common Symbols

DRIVER MODEL (IDM, DSM, etc.) PARAMETERS

v_d	desired driving velocity (in the absence of other constraints)
s_{min}	minimum allowable distance to lead vehicle
T_{min}	minimum allowable time headway to lead vehicle
a_{max}	maximum acceleration
b_{max}	maximum braking deceleration
Γ_{max}	maximum lateral acceleration during cornering
Δ	driver's cognitive safety margin for curvature estimation in corner
$iTTC_{min}$	minimum inverse time-to-collision
$iTTC_{max}$	maximum inverse time-to-collision

OPTIMAL CONTROL THEORY FOR ECO-DRIVING

J	objective (or 'cost') function
ϕ	terminal objective function
L_{idm}, L_{dsm}	stage cost functions associated with the IDM and DSM (respectively)
x, x_L	travelled distance of ego- and leading vehicles
s	inter-vehicle spacing
v, v_L	velocity of ego- and leading vehicles
a	acceleration of ego-vehicle
$s_d(v)$	desired inter-vehicle spacing (as a function of velocity)
$\kappa(x)$	road curvature (as a function of travelled distance)
t, t_0, t_f	time, start time, end time

VEHICLE DYNAMICS

m	vehicle mass
g	gravitational acceleration (approx. 9.81 m/s^2 on Earth's surface)
θ	road slope (expressed as an angle)
C_d, C_{rr}	coefficients of drag and rolling resistance
ρ	air density
A_f	frontal area of vehicle
F, F_t, F_b	total driving force, traction force, and brake force, respectively
F_{br}, F_{bf}	rear wheel and front wheel braking forces
T_m, ω_m	motor torque and speed
Q_b, i_b	battery charge and current
σ	battery state of charge
T_{max}, P_{max}	maximum torque and power

Preface

Addressing the reduction of energy usage and subsequent emissions of road vehicles, including carbon dioxide and nitrous oxides remains a key challenge faced by automotive engineers, researchers, and industry professionals. The findings illustrated in this book are the outcomes of three years of research, driven by the desire to actively contribute to this challenge. A key goal of this research was to develop a system that encourages drivers to adopt a more eco-friendly driving style, by interacting with the user in a simple and unobtrusive manner. The outcome is an Eco-Driving Assistance System (EDAS) that nudges the driver into appropriate behaviours rather than prescribing them, improves situational awareness without increasing workload, and trades-off energy efficiency with driver preferences in order to increase acceptance. This book illustrates the methods that have been employed in the process, which mix control engineering and human factors techniques in a synergic way. Moreover, the implementation and validation of a proof of concept of such a system is described. Across the different chapters, the reader will also find practical advice on how to develop and test an Eco-Driving Assistance System, which, as well as being useful for researchers, is particularly relevant for those working in the automotive industry.

A core principle of this book is the value of working in collaborative teams, whereby multidisciplinary expertise is embraced to achieve a common goal. When considering truly energy optimal driving, from a pure engineering perspective, the driving style that minimises energy consumption rarely matches typical driver preferences on acceleration, deceleration, cornering speed, and following distance. Understanding these driver preferences and the needs of end-users ensures that solutions reflect not only good engineering, but also an understanding of the human agents, to enable and champion long-term uptake of better driving habits and effective use of the technology. As a consequence, this book not only applies control engineering methodologies such as nonlinear Model Predictive Control (MPC) to the eco-driving problem, but also incorporates the human factors methods of Cognitive Work Analysis and Design with Intent to understand the task of fuel-efficient driving and develop subsequent interface prototypes holistically, including the human driver as part of the system. These disciplines combine to enable the successful development and testing of an operational EDAS system both within a simulator and on the road.

The book is organised as follows. Chapter 1 presents the concept of eco-driving, with a focus on the long-term maintenance of these behaviours, including training programmes and feedback devices. Chapter 2 documents the development of a complete Cognitive Work Analysis (CWA) to support environmentally energy-efficient driving. Chapter 3 uses data collected in naturalistic driving conditions for assessing the performance of some artificial driver models in terms of prediction or reproduction of driver speed choice during car-following and when cornering. Chapters 4 and 5 present a proof of concept of the Design with Intent (DwI) toolkit, which assists designers in creating novel designs and interfaces and builds on the CWA illustrated in Chapter 2. Chapter 6 illustrates a method for incorporating driver preferences, such as those on speed and following distance, into an eco-driving assistance system that trades-off between energy usage and driver satisfaction; the application of this method to a battery electric vehicle is discussed in Chapter 7. Chapter 8 discusses the efficacy of an EDAS in supporting the driver to reduce fuel consumption, which has been assessed in a simulator study. The question of whether EDAS benefits are appreciable for different types of propulsion technologies is debated in Chapter 9, where energy savings are evaluated for three different powertrains: a combustion engine-driven vehicle, a parallel hybrid electric vehicle, and a battery electric vehicle. Our research journey culminates in Chapter 10, where the implementation and on-road testing of a real-time predictive eco-driving assistance in a demonstrator vehicle is illustrated and compared to the simulator study. Finally, Chapter 11 summarises the lessons we learnt in our project and provides a number of practical suggestions for the development and testing of EDAS.

About the Authors

Dr Craig K. Allison is currently a Lecturer in Psychology at Solent University, Southampton. Craig earned his PhD in Web Science (Psychology) in 2016, his MSc in Web Science in 2011, and his BSc in Psychology in 2009, all from the University of Southampton. Craig's research background focused on spatial psychology, exploring how individuals used, explored, and maintained orientation within immersive 3D worlds. Craig transitioned to Human Factors research in 2015, during his time as a post-doctoral research fellow. Craig has worked on numerous topics within the Human Factors domain, primarily related to aviation and automotive industries. With expertise in both qualitative and quantitative analysis, Craig has extensive experience running research trials and working in multidisciplinary teams. Craig research interests primarily pertain to behaviour change relating to sustainability and the use and integration of novel technology within people's day-to-day lives.

Dr James M. Fleming earned his MEng and DPhil in Engineering Science from the University of Oxford in 2012 and 2016 respectively, following which he spent three years as a Research Fellow at the University of Southampton before joining the Wolfson School of Mechanical, Electrical and Manufacturing Engineering at Loughborough University in September 2019. He has research interests in the theory and practice of optimal control and model predictive control, with applications including energy-efficient autonomous driving, motorcycle stabilisation, and control of variable-speed wind turbines.

Dr Xingda Yan earned his BEng degree in automation from the Harbin Institute of Technology, Harbin, China, in 2012, and the PhD degree in Electrical Engineering from the University of Southampton, Southampton, UK, in 2017. He was a Research Fellow with the Mechanical Engineering Department, University of Southampton, Southampton, UK. Xingda is currently an automotive power engineer at Compound Semiconductor Applications Catapult, Newport, UK, and a visiting researcher with the Mechatronics Engineering Group, University of Southampton. His research interests include power electronics, hybrid system modelling and control, model predictive control, hybrid electric vehicle modelling, and energy management.

Dr Roberto Lot earned his Master's Degree in Mechanical Engineering and his PhD in Mechanics of Machines. He started his academic career at the University of Padova (Italy) as Assistant Professor in 1999 and became Associate Professor in 2005. In 2014 he was appointed to the chair of Automotive Engineering at the University of Southampton (UK), then in 2019, he moved back to Padova. His research focuses on the automotive sector and specifically on the dynamics and control of road and race vehicles. In 20 years of activity, he has directed several national and international research projects contributing to making our vehicles safer, faster, and more eco-friendly and publishing more than 120 scientific papers.

Professor Neville A Stanton, PhD, DSc, is a Chartered Psychologist, Chartered Ergonomist, and Chartered Engineer. He has recently retired from the chair in Human Factors Engineering in the Faculty of Engineering and the Environment at the University of Southampton in the UK. He has degrees in Occupational Psychology, Applied Psychology, and Human Factors Engineering and has worked at the Universities of Aston, Brunel, Cornell, and MIT. His research interests include modelling, predicting, analysing, and evaluating human performance in systems as well as designing the interfaces and interaction between humans and technology. Professor Stanton has worked on design of automobiles, aircraft, ships, and control rooms over the past 30 years, on a variety of automation projects. He has published 50 books and more than 400 journal papers on Ergonomics and Human Factors. In 1998 he was presented with the Institution of Electrical Engineers Divisional Premium Award for research into System Safety. The Institute of Ergonomics and Human Factors in the UK awarded him the Otto Edholm Medal in 2001, the President's Medal in 2008 and 2018, the Sir Frederic Bartlett Medal in 2012, and the William Floyd Medal in 2019 for his contributions to basic and applied ergonomics research. The Royal Aeronautical Society awarded him and his colleagues the Hodgson Prize in 2006 for research on design-induced, flight-deck error published in the *Aeronautical Journal*. The University of Southampton has awarded him a Doctor of Science in 2014 for his sustained contribution to the development and validation of Human Factors methods.

Acknowledgements

We want to acknowledge the financial support of the Engineering and Physical Sciences Research Council (EPSRC), the primary funder for the research work undertaken, under grant EP/N022262/1 "Green Adaptive Control for Future Interconnected Vehicles".

The authors give thanks to all those who participated in elements of this research, for instance attendees at our design and dissemination workshops, and participants of the simulator study at the University of Southampton, as without their time this work would not have been possible.

List of Abbreviations

CWA	Cognitive Work Analysis
CAT	Contextual Activity Template
ConTA	Control Task Analysis
SRK	Skills, Rules, and Knowledge
SOCA	Social Organisation and Cooperation Analysis
StrA	Strategies Analysis
WDA	Work Domain Analysis
WCA	Worker Competencies Analysis
SBB	Skill Based Behaviour
RBB	Rules Based Behaviour
KBB	Knowledge Based Behaviour
DwI	Design with Intent
EPSRC	Engineering and Physical Sciences Research Council
IEEE	Institution of Electrical and Electronics Engineers
ADAM	Automotive Data Acquisition Module
ADAS	Advanced Driver Assistance System(s)
IDM	Intelligent Driver Model
TTC	Time to Collision
TLC	Time to Lane Crossing
GPS	Global Positioning System
HMI	Human-Machine Interface
MPC	Model Predictive Control
DSM	Driver Satisfaction Model
RMS, RMSE	Root Mean Square, Root Mean Squared Error
MAXE	Maximum Error
EV	Electric Vehicle
HEV	Hybrid Electric Vehicle
PHEV	Plug in Hybrid Electric Vehicle
DC	Direct Current
AC	Alternating Current
SoC	State of Charge
ICE	Internal Combustion Engine
ECMS	Equivalent Consumption Minimisation Strategy
RPM	Revolutions per Minute
LCD	Liquid Crystal Display
GUI	Graphical User Interface
PC	Personal Computer
TCP/IP	Transmission Control Protocol / Internet Protocol
ANOVA	Analysis of Variance
ANCOVA	Analysis of Covariance
TLX	Task Load Index
OCP	Optimal Control Problem

GHG	Greenhouse Gases
CO_2	Carbon Dioxide
NOx	Nitrous Oxides
ACC	Adaptive Cruise Control
AAP	Active Accelerator Pedals
EU	European Union
US	United States of America
UK	United Kingdom

1 Eco-Driving
Reducing Emissions from Everyday Driving Behaviours

INTRODUCTION

There is growing acknowledgement of the role of man-made emissions in climate change (Thornton & Covington, 2015). Despite accepting the role of anthropometric generated greenhouse gases (GHGs) including carbon dioxide (CO_2) and nitrous oxides (NO_x) in changing global temperature, limited concerted steps to combat the carbon footprints of individuals have been taken. Whilst it can be difficult for an individual to consider the impact of their personal energy use, Vandenbergh et al. (2007) argue that approximately 30–40% of total GHG emissions can be directly attributed to individual energy needs required to support 21st-century lifestyles. Dietz et al. (2009) propose that emissions could be significantly reduced, with no consequence to lifestyle, if individuals simply acted in an eco-friendlier manner, taking steps to reduce excessive and waste energy consumption. This chapter presents a narrative review, aligned from a human factors perspective, examining the steps taken within transportation to reduce emissions, with a particular focus on the concept of eco-driving.

Whilst future chapters will more fully consider the influence of vehicle mechanics and drivetrain optimisation on fuel efficiency, this chapter will introduce the topic of eco-driving, the behavioural approach to reducing the overall vehicle fuel use, while examining the role of training and feedback in supporting eco-driving behaviours in the long term. Despite evidence demonstrating the initial effectiveness of eco-driving courses (Kurani et al., 2015), these benefits are often short term, with drivers rapidly returning to previous habits (Wåhlberg, 2007). Research has, however, demonstrated that regular feedback, such as provided by in-vehicle interfaces, can be an effective tool in helping drivers maintain eco-driving behaviours (Meschtscherjakov et al., 2009), although ambiguity remains regarding the best sensory modality to present eco-driving feedback (McIlroy et al., 2016). Despite uncertainty regarding the best way to support eco-driving behaviours, it is apparent that their adoption can lead to both reduced GHG emissions (Alam & McNabola, 2014) benefiting the environment and allowing financial savings (Beusen et al., 2009) benefiting the driver.

TRANSPORTATION CONTRIBUTION TO GHG

A key component of personal energy consumption, and a significant source of GHG emissions, is transport (Fuglestvedt et al., 2010). Hill et al. (2012) posit that transport is the second largest source of GHG emissions in the European Union (EU), after electricity generation. Of this, road transportation, specifically automobiles, is the biggest contributor, accounting for approximately 75% of the total transport-related GHG emissions. Similarly, considering Japan, it has been estimated that automobiles account for approximately 87% of transportation-related CO_2 emissions (Ando & Nishihori, 2012). Hill et al. calculate that car transportation equates to approximately 20% of the total CO_2 EU emissions across all sectors. In addition to being a significant contributor to GHG emissions, transport is one of the few sectors where GHG emissions have grown in recent decades. Between 1990 and 2009, there was a 24% decrease in total GHG emissions across the EU; transportation GHG emissions, in contrast, rose by 29% (Hill et al., 2012). Despite increased contributions from both maritime and aviation, which was the largest growth transport sector, automotive transportation played a significant role in the increasing quantity of emissions. The trend for significant transport emissions is, however, a global problem. In the United States of America (USA), personal transportation accounts for approximately 32%–41% of total CO_2 emissions (Bin & Dowlatabadi, 2005; Vandenbergh & Steinemann, 2007). The US transportation-related emissions exceed emissions from the US industrial sector and are greater than the total emissions from all other individual countries, with the exception of China (Boden et al., 2009), which is currently the world's leader in GHG and CO_2 emissions (Gregg et al., 2008; Netherlands Environmental Agency, 2017). When considered as a whole, Barkenbus (2010) estimates that approximately 8% of the world's total CO_2 emissions are a result of transportation, primarily related to the use of automobiles.

The idea that automobile use has a significant and long-term negative impact on our environment is not novel, with Berntsen and Fuglestvedt (2008) estimating that, based on emissions figures from the year 2000, automobile use has approximately four times greater global warming effect than aviation. Extending this analysis, Skeie et al. (2009) estimate that approximately 14% of global mean temperature change will be a direct consequence of transportation, with automobiles being the primary contributor. Seeking ways to minimise transportation, and specifically automobile related, emissions should therefore be considered a priority for helping combat climate change. Due to the scale of transportation GHG emissions, small-scale savings made within this sector can act as a key cornerstone for emission reduction, presenting a prime candidate for collective action.

ECO-DRIVING

Whilst it is fundamentally clear that the most dramatic way to reduce transportation-related GHG emissions is to not take car journeys, an option ingrained into some newer drivers during training (Strömberg et al., 2015), Barkenbus (2010) proposed that GHG emissions can be significantly reduced by altering driving style. Barkenbus argued that should individuals adopt a measured driving style, referred to as eco-driving, fuel use, and as a consequence GHG emissions, could be dramatically reduced. Adopting eco-driving practices has been demonstrated to hold considerable emission reduction and fuel saving potential, with previous research suggesting that emissions could be reduced by 5%–20% (Stillwater & Kurani, 2013) with fuel usage being reduced by between 5% and 10% (Martin et al., 2012). Barkenbus (2010) proposes that eco-driving is characterised by behaviours such as modest acceleration, early gear changes, limiting the engine to approximately 2,500 revolutions per minute (RPM), anticipating traffic flow to minimise braking, driving below the speed limit, and limiting unnecessary idling. The use of these techniques, and subsequent emissions and fuel use reductions, can be achieved without significantly increasing journey time (Barth & Boriboonsomsin, 2009). Although Barkenbus notes that eco-driving best practices include non-driving behaviours, for example, ensuring that the car is well serviced and making appropriate navigation decisions, Sivak and Schoettle (2012) posit that eco-driving has primarily been considered in terms of driving style following vehicle selection. Whilst Barkenbus concentrates heavily on the operational decisions a driver makes during a journey, Sivak and Schoettle (2012) posit that the concept of eco-driving can be extended to also include strategic and tactical decisions that drivers make. Strategic decisions relate to initial vehicle selection, for example choosing a vehicle that maximises fuel economy as well as ongoing vehicle maintenance, for example ensuring tyres are adequately inflated. Conversely, tactical decisions relate to gestalt decisions about the overall journey being undertaken such as navigational choices, for example changing route in order to avoid traffic and decisions related to current vehicle load. The adoption of eco-driving behaviours is advantageous in that it is an emission-reduction technique that is available to all drivers. Whilst replacing older vehicles, typically characterised by inefficient internal combustion engines, with more technologically capable, environmentally friendly, and fuel-efficient drivetrains is desirable, for example the use of electric, hydrogen fuel cell, or hybrid vehicles (Chan, 2007; Lorf et al., 2013), or the use of alternative, cleaner fuel sources, such as compressed natural gas (CNG) (Windecker & Ruder, 2013), the initial financial investment required for such vehicles means that, for many, this is an untenable option. Eco-driving, as a behavioural intervention, in contrast, presents an opportunity for all drivers, regardless of circumstance, to reduce the environmental impact of their own personal transportation.

As previously stated, in addition to the environmental benefits of eco-driving, adopting these behaviours has direct financial benefits to the driver, primarily due to reduced fuel consumption and reduced fuel costs per journey (Alam & McNabola, 2014; Mensing et al., 2014). Kurani et al. (2015) reviewed 32 studies that considered fuel economy and the uptake of eco-driving behaviours. They found, on average, a 9% fuel saving as a consequence of adopting eco-driving techniques, although they note that this figure varies considerably, with in-simulator studies typically recording greater potential savings than on-road trials. Due to the financial benefits associated with reduced fuel consumption, the potential benefits of eco-driving are increasingly compelling, especially considering rising fuel prices (Heyes et al., 2015). Beusen et al. (2009) note that a 5% fuel saving for a typical driver, a figure equal to the lower estimate from Kurani et al. (2015), would result in approximately £250 (≈ $350) a year financial saving.

ECO-DRIVING KNOWLEDGE

Despite the environmental and financial benefits of adopting eco-driving, understanding the best way to disseminate this information to drivers remains elusive. This task is further complicated by significant variation in driver knowledge regarding eco-driving behaviours. Strömberg et al. (2015) compared the eco-driving knowledge of nine older and nine younger drivers in Sweden. They found that whilst the term eco-driving was known to all bar two experienced drivers, the level of knowledge relating to eco-driving varied greatly both within and between groups. Unsurprisingly, as eco-driving is increasingly normalised within newer driving education, younger drivers had similar awareness and depth of knowledge regarding the eco-driving concept. Interestingly, however, the quality of the eco-driving training varied considerably between individuals, with some having been taught eco-driving techniques as integral to driving, normalising the technique. Others, however, were taught that eco-driving was a series of techniques that could be used as the driver saw fit. This marked difference in the quality of driver training within a limited sample of just nine younger drivers suggests that greater efforts are needed to standardise eco-driving training. Whilst it is unfeasible to track the driving actions of drivers once they pass their test, mandating set driving styles during training may have a large-scale positive impact on future drivers' behaviours. Conversely, the more experienced drivers varied greatly in their level of awareness of the eco-driving concept, with one participant having been on several courses focussed on the topic. More experienced drivers had an awareness of the concept of eco-driving due to information passed from younger family members and colleagues, but lacked an in-depth understanding. Despite an absence of formal understanding, the potential of eco-driving was generally well received among participants.

Seeking to understand the level of eco-driving knowledge held by members of the general public, McIlroy and Stanton (2017) explored participants' awareness of eco-driving techniques, using an online questionnaire. McIlroy and Stanton (2017) found that of the 321 respondents, 81% were able to demonstrate knowledge of at least one eco-driving behaviour, for example ensuring tyres are adequately inflated and/or minimising the engine revolutions. Although identifying that the majority of respondents were aware that their actions could influence fuel usage, significant variation regarding overall knowledge was apparent within the sample, with participants displaying limited detailed knowledge of specific techniques. McIlroy and Stanton note, however, that there was a general correspondence between the techniques suggested by the sample and techniques suggested within academic literature to reduce fuel use, for example those provided by Barkenbus (2010) and Hooker (1988). The general lack of detailed knowledge possessed by respondents suggests that whilst eco-driving is often viewed as a cost-effective way to reduce fuel use (Birol, 2017), requiring no financial investment (Saboohi & Farzaneh, 2009), significant financial investment may be needed by governments and charitable organisations to generate greater driver awareness and knowledge. Equally, drivers themselves may be required to invest a considerable amount of time and effort into developing and maintaining the skills required to see a benefit of eco-driving practices, with studies reporting that some behaviours, including maintaining low RPM and early gear changes are difficult to sustain long term (Delhomme et al., 2013).

One factor that does constrain the dissemination of eco-driving knowledge to the wider public is the lack of established guidelines on how best to achieve eco-driving. Kurani et al. (2015) argue that despite eco-driving behaviours being well known, for example avoiding excessive use of a vehicle's air-conditioning system, these behaviours are often generic and imprecise. Eco-driving can, therefore, be best understood when considered in terms of qualitative behaviours rather than quantitative actions. For example, gentle acceleration is a key concept within the eco-driving framework, however what is meant by gentle acceleration can vary significantly between individuals. Because of the lack of consistent and established quantified eco-driving guidelines, it is difficult to directly compare the multiple research studies that populate the literature. The current lack of consistent definitions may also partially explain the dramatic differences recorded within studies exploring the potential benefits of eco-driving, both in terms of overall fuel usage and total emissions. Barkenbus (2010) notes that the achievable fuel use savings documented within the literature as a consequence of adopting eco-driving ranges from 5% to 20%. Dula and Geller (2003) argue that a similar ambiguity exists in literature related to the converse of eco-driving, that of aggressive driving. They also argue that the lack of consistent definitions for aggressive driving makes comparisons between studies difficult, although are supportive of the view that aggressive driving is associated with greater fuel use. Sivak and Schoettle (2012) note that continually driving in an aggressive way, for example with maximum use of the accelerator pedal and heavy braking, with high vehicle loads and use of ancillary vehicular systems, such as air-conditioning can increase fuel consumption by approximately 45%. Although accurately estimating the potential benefits and fuel savings as a consequence of adopting eco-driving behaviours is a difficult task, it is clear that driving style does impact total fuel use and subsequent GHG emissions.

Provision of eco-driving advice is further constrained and complicated by the increasing availability of alternative vehicle drivetrains. Traditionally, eco-driving research has focussed on internal combustion engines (ICEs), however it is clear that drivetrains such as hybrid electric vehicles (HEVs), such as the Toyota Prius and plug-in electric vehicles (PHEVs), such as the Nissan Leaf and Mitsubishi Outlander PHEV, with regenerative braking (Shabbir & Evangelou, 2014) are increasingly available. HEVs and PHEVs provide a challenge for eco-driving information dissemination as their energy dynamics, the way they consume, use and regenerate energy, differs from traditional ICE vehicles (McIlroy et al., 2014), meaning that advice given to owners of these vehicles may need to be specifically tailored. To examine eco-driving techniques of HEV owners, Franke et al. (2016) interviewed 39 Toyota Prius drivers, who regularly logged their fuel use online and achieved higher than average efficiency. Franke et al. found that participants held a diverse level of understanding regarding their HEV, including many holding false beliefs in relation to the energy recuperation system and when best to utilise the onboard electric engine. Franke et al. note that even within their relatively small sample of fuel-efficient drivers, significant variations in eco-driving strategies were observed, identifying clear differences in the level of knowledge held by the driving population. This finding suggests that there is a need for clearer guidelines to be presented to HEV drivers. Of central importance, however, was that many strategies employed by participants were the same as would emerge when considering ICE vehicles, including anticipation of traffic flow and limiting overall speed on high-speed roads such as autobahns. This suggests that HEV eco-driving strategies build on those already employed within ICE vehicles.

ECO-DRIVING TRAINING

One approach to the dissemination of eco-driving information is through prescribed training courses, a technique that has been previously pursued. In a simple comparison between eco-driving advice and previous eco-driving training, Andrieu and Saint Pierre (2012) explored the fuel use of two sets of French drivers. They found that despite fuel use reduction for both groups being highly comparable, approximately 12% saving, the training group displayed positive behaviours more frequently and consistently, leading to the conclusion that whilst informing users about eco-driving is worthwhile, as consistent performance was better following the eco-driving course. One weakness of this work, however, was that the samples used were limited and only based on two drives. Exploring the long-term impact of eco-driving, Beusen et al. (2009) monitored the fuel use of 10 drivers for a year, and explored the impact of an eco-driving course that participants undertook 6 months into the monitoring period. Throughout the year, vehicular data, including fuel use, was collected from an in-vehicle controller area network (CAN)-bus sensor. Beusen et al. found that after the course, participants' fuel consumption was reduced by an average 5.8%, signifying the effectiveness of the training course. Additionally, Beusen et al. found that this effect remained 6 months later at the end of the monitoring period, leading to the suggestion that the savings induced as a consequence of the course were permanent. Following concerns over the validity of this finding, the data collected by Beusen et al. was reanalysed by Degraeuwe and Beusen (2013), with an additional factor added to the analysis, ambient temperature. Ambient temperature is inversely correlated with fuel economy, as temperature rises, such as occurs during summer months, fuel consumption falls. As the training course was typically offered to drivers in March and June, ambient temperature was rising. Upon reanalysis, accounting for this effect, it was found that although the training course did significantly reduce fuel use upon initial training, this effect was not permanent and disappeared once the effect of changes within ambient temperature was accounted for.

Despite evidence highlighting the benefits of eco-driving training, in terms of both fuel use and emission generation, studies have also indicated that the value of eco-driving education deteriorates over time. Geiler and Kerwien (2008, cited in Heyes et al., 2015) explored the impact of an eco-driving training course and found that despite an initial fuel saving of 7% after the course, in a follow-up examination 10 months later, this saving had dropped to 4%. Wåhlberg (2007) reported similar deterioration in the impact of eco-driving training, specifically examining bus drivers. Wåhlberg found that despite the initial success of an eco-driving training course, as indicated by a 6% drop in fuel use, during a follow-up investigation 12 months later, it was found that the fuel saving had been reduced to 2%. Likewise, Zarkadoula et al. (2007), also investigating bus drivers, found that following an eco-driving training course, overall fuel usage dropped by 10.2%. Zarkadoula notes, however, this figure had dropped to a 4.35% saving compared to initial fuel use just 2 months after the course. Comparable declining results following an eco-driving intervention were reported by Lai (2015), exploring financially incentivised bus drivers. Although drivers received a monetary bonus for achieving greater fuel efficiency, it was found that the impact of the eco-driving course rapidly declined, even though the incentive scheme remained active.

These studies suggest that whilst eco-driving courses do promote fuel-efficient driving and can induce a long-term fuel saving, it is apparent that individuals reintroduce habitual non-eco-friendly driving behaviours following the course. Although an overall reduction in fuel usage remains, it is clear that these are much lower than would be expected based on impact of the initial training. These findings suggest, therefore, that training is insufficient to produce permanent changes in behaviour.

FEEDBACK AND ECO-DRIVING

Providing drivers feedback regarding their current driving and educating them about alternative driving styles has been seen as essential in eliciting long-term changes in behaviour (Tulusan et al., 2012). It may be that many drivers would adopt an eco-driving approach if they had greater awareness and understanding of the impact of their current behaviour (Abrahamse et al., 2005). Froehlich et al. (2009) describe eco-feedback technologies as *"technology that provides feedback on individual or group behaviours with a goal of reducing environmental impact"*. Abrahamse et al. (2005), in a literature review of personal energy use, identified a variety of factors that influence motivation to engage in eco-friendly behaviours. It was noted that whilst commitment was a key factor, feedback regarding current behaviours was an important variable of note. Specifically examining transportation, Froehlich et al. (2009) and Meschtscherjakov et al. (2009) argue that providing feedback is a cost efficient way to encourage and reinforce eco-driving practices.

The impact of driver feedback has been demonstrated across a variety of situations and research studies. Lauper et al. (2015) following a longitudinal questionnaire study argued that feedback is essential in maintaining drivers' ability to regulate their behaviour and encourage future behavioural implementations. Tulusan et al. (2012) explored the impact of eco-driving feedback on the fuel use of corporate drivers, comparing the fuel efficiency of two independent groups of drivers, either with or without access to the smartphone application DriveGain. DriveGain provided live feedback to the driver, including appropriate time to change gear, appropriate speed for the road and route, and a measure of appropriate braking. Encouraging eco-driving behaviours within corporate drivers is an important initiative as whilst the average private car drivers travel 8,500 miles a year, corporate drivers travel approximately 21,500 miles (Tulusan et al., 2012). Participants had access to the application for eight weeks but were free to choose when they wished to use the application. Use of the application was not enforced within the study to encourage more naturalistic results. It was found that of the participants who had access to the application, fuel economy improved by approximately 3%, compared to the control group. These findings are in line with other studies exploring the benefits of eco-driving schemes (Geiler & Kerwien, 2008, cited in Heyes et al., 2015; Boriboonsomsin et al., 2010; Mensing et al., 2014), despite investigating a population of drivers who would not financially benefit from changing their driving style. Importantly, this study offers clear indication that drivers can adopt eco-driving, without the need for financial incentives when provided with feedback.

Evidence that eco-driving behaviours can be encouraged without financial incentives suggests that drivers have motivations to adopt more environmentally friendly behaviours beyond financial. Dogan et al. (2014) argue that due to the relatively small economic savings incurred as a result of eco-driving, drivers may consider the required effort to outweigh the benefit. Dogan et al. argue, however, that the environmental saving message could overcome this perceived effortful behaviour due to a fundamental motivation to make a positive impact. Dogan et al. draw upon the work of Heyman and Ariely (2004) and suggest that fundamental differences in motivation, such as desires for a financial reward, or a desire to have a positive impact on society and the environment, can be considered to encourage eco-driving behaviours. Dogan et al. presented an eco-driving questionnaire to 305 respondents, who received either financial, environmental, or no information about the impact of a series of driving scenarios, for example waiting at a railway crossing and driving with inadequately inflated tyres. When considering the value of eco-driving behaviours, it was found that respondents presented with environmental feedback viewed the benefits of adopting eco-driving behaviours as significantly greater than respondents presented with financial or no information. Similarly, respondents presented with environmental information gave greater indication that they would be willing to change their current behaviour in response to the feedback than participants presented with financial information. Although both types of feedback were effective at encouraging eco-driving responses, Dogan et al. noted that environmental feedback resulted in stronger intentions to change. Taken together with the work of Tulusan et al. (2012), it is clear that the intention to make a positive environmental impact can be sufficient to motivate the intention for behavioural change, beyond financial gain. This suggests that a variety of motivations should be addressed when considering ways to encourage the adoption of eco-driving. Central to encouraging this behaviour change, however, is the adequate provision of feedback.

As discussed previously (Tulusan et al., 2012), the use of in-vehicle visual feedback, provided by a smartphone application, has been demonstrated to be effective in encouraging the uptake of eco-driving behaviours. In this way, eco-driving feedback can be provided with minimal initial expense, and does not require extensive vehicle modifications. Visual feedback can be a valuable, as well as a low-cost approach to encouraging eco-driving behaviours (Froehlich et al., 2009; Meschtscherjakov et al., 2009; Tulusan et al., 2012). Evidencing the value in visual displays for eco-driving feedback, Van der Voort et al. (2001) presented participants in a driving simulator with visual feedback regarding their current driving style. They found that provision of visual feedback was able to reduce fuel use by 16% compared to everyday driving. Interestingly, however, they also found that just asking participants to drive in a fuel-efficient way lead to a 9% fuel saving, suggesting that active reminders can dramatically reduce fuel use. Building on these findings, it is unsurprising that visual feedback regarding driving style is increasingly built into newer, high specification, production automobiles. As an example, Honda has begun to implement a speedometer that changes colour based on vehicle performance and the rate of acceleration (Azzi et al., 2011). Visual feedback is advantageous in that it can often be more detailed than haptic or auditory information, and can be further enhanced via the use of symbols and agreed upon colour coding, for example, the idea that green is associated with positive and red associated with negative outcomes is a cross-cultural standard (Madden et al., 2000). Due to the potential of different symbols, it should be possible to provide personalised feedback in order to target effective motivational messages to different individuals based on whether they are monetarily or socially motivated (Heyman & Ariely, 2004; Dogan et al., 2014;). Young et al. (2009) developed a visual feedback interface to encourage the uptake of eco-driving behaviours. The interface, which provided visual feedback was developed following a combination of Cognitive Work Analysis (Rasmussen, 1986), applied to eco-driving (Birrell et al., 2012) and user questionnaire feedback. Despite Young et al. (2009) identifying significant development opportunities following user tests, it was clearly seen that visual feedback benefited participants by providing information relating to current driving style and providing useful eco-driving information.

Despite the advantages offered by visual eco-driving feedback, a fundamental limitation of visual feedback is that it can be inherently distracting (Azzi et al., 2011). As a consequence of diverting drivers' attention from driving, visual feedback devices can potentially negatively impact driver safety, for example by impairing their ability to detect potential hazards (Recarte & Nunes, 2003). A driver cannot concurrently examine both the ever-changing road situation and the visual feedback (Young et al., 2007). This is true even for head-up displays (HUDs) whereby the driver does not have to actively look away from the road. Issues relating to information focus, whereby the user examines the displayed information rather than the dynamic environment in which they are travelling has been well documented in other transportation sectors with a history of HUD use, for example aviation (Crawford & Neal, 2006). Specifically, within the automotive domain, Summala et al. (1998) found that attending to panel displays significantly impaired driving performance and reaction times. This is especially apt when considering travelling at higher speeds, for example during motorway, freeway, and autobahn journeys and could have significant safety implications.

Kircher et al. (2014) argue that in order to be an effective tool and modify behaviour, feedback must be inherently distracting. When this positioning is accepted, the issue is less whether feedback is distracting, but rather becomes how and when feedback information should be made available to the driver. Kircher et al. investigated, within a driving simulator study, the time drivers examined a visual feedback device and found, as would be anticipated, that participants spent significantly longer examining the device when feedback was shown, and largely ignored the device when no feedback information was displayed. They note, however, that the time drivers spent examining the feedback device may be artificially inflated in a driving simulator study compared to the real world. Participants are aware of the study, are operating within an artificial environment, with significantly reduced risks and do not have access to other potential distractors which they may have in the real world, including the radio, other onboard systems, and access to refreshments. Kircher et al. suggest that research is needed to identify when the most appropriate time would be to provide feedback to the drivers, for example under what traffic conditions, and at which navigational points of the journey to both inform the driver and minimise potential negative distraction effects. Davidsson and Alm (2014) note, however, that this is not a simple task as different drivers have different needs based upon their current driving context and previous experience, and as such feedback systems need to not only adapt to drivers' current actions but also their long-term needs. Rouzikhah et al.

(2013) explored the impact of visual eco-driving feedback, as presented in text form, on participant perceived workload, during a series simulated drives. Rouzikhah et al. (2013) compared baseline driving, with no secondary task, driving with access to eco-driving feedback, driving whilst changing a CD, and driving whilst being required to enter navigation information. Rouzikhah et al. (2013) found that although there were significant differences in participants' perceived workload between changing a CD and entering navigational information compared to the baseline task, participants did not report an increased workload when examining the eco-driving feedback. Rouzikhah et al. note that although both changing a CD and entering navigational information tasks were more physically demanding, leading to increased physical workload within these scenarios, the eco-driving message should have increased cognitive demand. Examining the data generated within the simulator revealed, however, that participants did not appear to alter their driving style following provision of the eco-driving feedback, suggesting that although the eco-driving message was not distracting, it was not cognitively engaging enough to modify behaviour. This finding supports the work of Kircher et al. (2014) who suggest that feedback must be inherently distracting to encourage behavioural modification.

An alternative use of real-time visual feedback is to offer post journey advice. In addition to providing drivers useful real time eco-driving advice, smartphone applications can be used to provide drivers with a detailed breakdown of their journey statistics, including fuel use and emissions. This technique gives drivers access to extensive post journey feedback, which can be compared across multiple journeys. Husnjak et al. (2015) used a smartphone application to measure fuel use and CO_2 emissions for both a standard drive and a drive whereby eco-driving techniques were employed. Although the application did not provide in-journey advice, Husnjak et al. found that, on a single repeated route, fuel consumption was reduced by 23% and CO_2 emissions were reduced by 31% as a consequence of seeing information related to the previous journey. By considering this improvement, it may be that encouraging awareness of the presence of the feedback device results in a social facilitation effect (Allport, 1924) whereby drivers feel under observation and hence change their behaviour. The long-term facilitative impact of these devices has however been questioned. Rolim et al. (2016) compared the change in performance of a group of 40 drivers, who either had or had not received weekly feedback relating to their driving style and fuel efficiency following extensive monitoring. They found that rather than improving drivers' fuel economy, the feedback reports reduced fuel efficiency, with the drivers displaying a higher number of accelerations and a greater number of overall journeys, especially shorter journeys of less than 2 kilometres. Rolim et al. (2016) found that although negative feedback would trigger subsequent improvement in driver behaviour, the opposite was true when participants received positive feedback, with drivers failing to maintain their improvements and reverting to previous habits. This suggests that rather than just providing feedback, drivers must have a desire to achieve maximum fuel economy from their vehicle. Building on this approach, multiple online communities have developed, for example ecomodder.com and fuelly.com which actively encourage members to share their current fuel economy statistics and techniques with others, actively encouraging competition among members, and fostering social facilitation.

Eco-Driving: Reducing Emissions from Everyday Driving Behaviours 11

ECO-DRIVING FEEDBACK IN OTHER SENSORY MODALITIES

Despite the potential of visual feedback, eco-driving feedback has been successfully provisioned in other physical modalities, primarily haptic or touch-based feedback. Building on Wickens' (2002; 2008) multiple resources theory, Van Erp and Padmos (2003) suggest that driving is fundamentally a visual task, and note that providing feedback in the form of other sensory modalities may be informative enough to impact behaviour but not sufficiently distracting to reduce performance. Haptic feedback within automotive vehicles is most commonly realised by applying a resistive force to the accelerator pedal. Haptic feedback utilising the accelerator pedal has been trialled as both a form of speed management (Adell & Várhelyi, 2008) and as a method for encouraging eco-driving (Azzi et al., 2011). Larsson and Ericsson (2009) installed haptic force feedback into four postal delivery service vehicles. They found that the maximum acceleration force drivers applied to the pedals was significantly reduced. As accelerator force is a significant predictor of fuel usage and emission, this suggests that haptic feedback can be a valuable tool in supporting eco-driving. Adell and Várhelyi (2008) installed active accelerator pedals (AAPs) into 281 test vehicles to examine the impact of haptic feedback on speeding. The AAPs provided a counterforce to the drivers' effect of acceleration as they approached and exceeded the speed limit. AAPs were active within the installed vehicles for between six months and a year whenever the vehicle was within the designated test city and could not be turned off. Participants completed questionnaires at the start and end of the investigation, whereby demographic details were collected and their attitudes towards the feedback devices assessed. It was found that participants' opinions of the AAPs were largely positive with 79% rating the device as "Good" or "Very Good", however drivers' enjoyment of driving was reduced whilst using the AAPs, and drivers reported that individual journeys took longer to complete. Participants noted, however, that they had a greater awareness of safety and travelled at lower speeds during the study, suggesting that the AAP had achieved its primary goal of decreasing road speed. Interestingly, despite the positive self-reported results, it was found that only 35% of the participants would be willing to pay to keep the device once the study was over, suggesting that despite perceived benefits, generally users were not willing to financially invest in the system. Adell and Varhelyi note that the participants found the AAP *"useful but not satisfactory" (P50)* suggesting that hurdles to adoption were still present, as demonstrated by participants' lack of willingness to pay for the system. Nevertheless, this work does demonstrate the potential for haptic feedback in influencing drivers' behaviour.

The influence of a haptic feedback pedal has been examined within studies focussed on supporting eco-driving (Azzi et al., 2011; Birrell et al., 2013). Using an independent group design, Azzi et al., compared the efficiency of visual feedback, a haptic force pedal, and a combined system to examine the most effective way to provide eco-driving feedback to drivers, based on fuel usage and emissions. Using a driving simulator, participants were required to drive a set route within a traffic-free urban environment. It was found that participants provided with feedback recorded significantly lower emissions than participants assigned to a control group provided with no feedback. Limited differences were observed, however, as a consequence of the type of feedback offered, suggesting that the haptic force pedal was as successful at reducing emissions as the visual feedback displays, without many of the associated disadvantages, relating to split attention previously discussed. It was found that the combination of haptic and visual feedback resulted in significantly lower emissions than visual feedback alone, suggesting that the haptic feedback was highly salient to the driver. These results suggest that haptic feedback would be advantageous in encouraging eco-driving behaviours. Birrell et al. (2013) also exploring haptic feedback, explored the impact of a vibration alert when excessive force was applied to the accelerator pedal. Within a repeated measures simulator study, participants drove a set route within an urban and extra-urban environment, across three conditions. Participants were instructed to either drive normally, drive as fuel efficiently as possible, or provided with haptic feedback from the pedal if they produced excessive force on the accelerator, presented as a vibrational alert. Birrell et al. found that across the three conditions, average speed and journey time did not significantly differ. It was found that although average accelerator position did not change across the conditions, the maximum accelerator force applied within the conditions did significantly vary. Participants applied less maximum accelerator force when asked to drive fuel efficiently and applied a lower still maximum force when provided with haptic feedback. As fuel usage and emissions are heavily correlated with accelerator force, this finding suggests that just asking participants to follow an eco-driving approach can reduce fuel usage and subsequent emissions. Furthermore, this study suggests that haptic feedback can be used to positively encourage eco-driving behaviours with no significant negative consequences to journey statistics including time taken.

Despite these encouraging findings, haptic feedback does have several notable disadvantages that could reduce acceptance and use. Adell and Várhelyi (2008) identified cost as a significant barrier, as, unlike visual feedback devices, which are typically low-cost and easy to install, for example, via the use of a smartphone application, haptic devices must be engineered and retrofitted to a vehicle at comparatively considerable cost. This retrofitting is further complicated when considering that not all older automotive vehicles are suitable for the adaption, potentially reducing uptake further. A second key limitation is the impact of driver comfort and potential strain that could result from the use of haptic feedback pedals. Jamson et al. (2013) following a short simulator study suggest that over the course of prolonged journeys, drivers could experience fatigue and discomfort using a feedback pedal. Jamson et al. note that further research is needed to explore the long-term effect of haptic feedback on driver comfort and the potential for adaption to the feedback. This comment is especially telling considering Adell and Várhelyi (2008) previous finding that the use of a haptic feedback pedal was seen to decrease drivers recorded comfort and enjoyment with driving following long-term use. The short-term exposure and limited sample within these studies however restricts the impact of these findings beyond a variable of interest in future research.

In addition to visual and haptic feedback, auditory feedback has been used to support eco-driving. One study to explore the impact of auditory feedback was provided by Zhao et al. (2015) examined the role of generic and adaptive feedback, provided by a visual display and voice prompts in improving participants' fuel efficiency. Participants were required, within a driving simulator study, to drive a route three times, either with no feedback, with generic eco-driving tips, or with dynamic eco-driving tips, based on their previous actions. Zhao et al. found that participants achieved a 3.43% saving when presented with generic eco-driving reminders. When provided with dynamic feedback, however, the fuel saving rose to 5.45%. These findings suggest that despite savings being possible based solely on creating greater awareness of eco-driving, personalisation, and direct feedback based upon drivers' actual actions can be more effective. Despite the effectiveness of auditory feedback, research has primarily used this modality to support visual and haptic interfaces (Staubach et al., 2014).

As it has been seen, visual, auditory, and haptic sensory channels have been suggested as potential sources of feedback to encourage the uptake and maintenance of eco-driving behaviours. Although evaluation of each feedback method has been undertaken and comparison between two modalities has been considered, limited work has directly sought to compare the effectiveness of all three methods of feedback in a single study. This gap was considered by McIlroy et al. (2016). Within a simulator study, the effectiveness of visual, auditory, and vibrotactile haptic feedback was compared, both to normal driving behaviour and when the drivers were told explicitly to eco-drive. McIlroy et al. found that although simply asking participants to drive economically was sufficient to reduce fuel use, this effect was increased by the inclusion of the feedback devices. McIlroy et al. found that the increased fuel efficiency was largely a consequence of an increase in coasting behaviour, gently decelerating the vehicle over a greater distance reducing the need for braking, which emerged within the feedback trials, but had not been part of the initial repertoire of behaviours which participants used when told to drive economically, suggesting participants were unaware of this prior to feedback provision. Regarding the effectiveness of the different feedback manipulations, it was found that visual feedback was the least effective at promoting fuel-efficient driving. Limited differences were, however, observed between the auditory and haptic, with both reducing fuel usage by similar margins. It was found, however, that when considering participants' subjective ratings of the alternative feedback modalities, auditory feedback received the lowest ratings, suggesting that despite a similar effectiveness rating, the haptic feedback would be more readily accepted. Within a similar study considering multiple feedback modalities, Staubach et al. (2014) examined the impact of an eco-driving feedback device that utilised haptic and visual feedback to encourage greater coasting behaviours. Participants were required to complete four simulated drives with and without the feedback device, within both an urban (speed limit 50km/h) and rural (speed limit 70km/h) environment. Objective data was collected from the simulator relating to participants driving and subjective data was collected within a post scenario questionnaire. Staubach et al. found that, depending on scenario, a mean fuel saving of 15.9%–18.4% was achieved using the feedback device, primarily as a consequence of reduced stopping and greater coasting. Although a fuel saving was identified, it was found that the feedback device remained distracting, with participants spending time looking at the visual aspect of the display and not the road environment. In addition, it was found that participants often ignored the initial warning from the pedal, suggesting that the saving achieved could have been much higher had participants responded when prompted.

With rising evidence that the enactment of eco-driving behaviours can be of significant benefits both economically and environmentally, the question of how to encourage the long-term adoption of these behaviours is key. From the research presented, it is clear that eco-driving can be encouraged by making drivers more aware of both the environmental consequences of failing to engage in an eco-driving approach, and also the financial benefits of following eco-driving techniques. Providing drivers with feedback regarding their driving habits does, however, appear central to the adoption of eco-driving behaviours. Understanding the best way, in terms of both the most effective and least distracting, to present this feedback to drivers, for example via visual, haptic, or auditory feedback remains an open question (Allison & Stanton, 2019).

CONCLUSIONS

From the research reviewed, it is clear that both overall energy consumption and associated GHG emissions from personal transportation can be reduced with the uptake of eco-driving behaviours (Barkenbus, 2010; Sivak & Schoettle, 2012). Eco-driving involves a variety of behaviours including gentle acceleration, anticipating traffic, and reducing unnecessary idling. Eco-driving training however has been shown to have limited long-term impact, despite encouraging initial fuel reduction, with research (Wåhlberg, 2007; Zarkadoula et al., 2007) indicating that this effect is short-term with drivers rapidly returning to their habitual behaviours. To counter this return to habitual behaviours, it is important to consider both individual motivations to eco-drive, including financial and environmental as well as to provide feedback on current driving performance. Eco-driving feedback has been offered in a variety of modalities, including visual and haptic. Despite the advantages of feedback devices in promoting eco-driving behaviours, caution is warranted in their use due to the potential of driver distraction.

REFERENCES

Abrahamse, W., Steg, L., Vlek, C., & Rothengatter, T. (2005). A review of intervention studies aimed at household energy conservation. *Journal of Environmental Psychology*, 25(3), 273–291.

Adell, E., & Várhelyi, A. (2008). Driver comprehension and acceptance of the active accelerator pedal after long-term use. *Transportation Research Part F: Traffic Psychology and Behaviour*, 11(1), 37–51.

Alam, M. S., & McNabola, A. (2014). A critical review and assessment of eco-driving policy & technology: Benefits & limitations. *Transport Policy*, 35, 42–49.

Allison, C. K., & Stanton, N. A. (2019). Eco-driving: The role of feedback in reducing emissions from everyday driving behaviours. *Theoretical Issues in Ergonomics Science*, 20(2), 85–104.

Allport, F. H. (1924). The group fallacy in relation to social science. *The Journal of Abnormal Psychology and Social Psychology*, 19(1), 60.

Ando, R., & Nishihori, Y. (2012). A study on factors affecting the effective eco-driving. *Procedia-Social and Behavioral Sciences*, 54, 27–36.

Andrieu, C., & Saint Pierre, G. (2012). Comparing effects of eco-driving training and simple advices on driving behavior. *Procedia-Social and Behavioral Sciences*, 54, 211–220.

Azzi, S., Reymond, G., Mérienne, F., & Kemeny, A. (2011). Eco-driving performance assessment with in-car visual and haptic feedback assistance. *Journal of Computing and Information Science in Engineering*, *11*(4), 041005.

Barkenbus, J. N. (2010). Eco-driving: An overlooked climate change initiative. *Energy Policy*, *38*(2), 762–769.

Barth, M., & Boriboonsomsin, K. (2009). Energy and emissions impacts of a freeway-based dynamic eco-driving system. *Transportation Research Part D: Transport and Environment*, *14*(6), 400–410.

Berntsen, T., & Fuglestvedt, J. (2008). Global temperature responses to current emissions from the transport sectors. *Proceedings of the National Academy of Sciences*, *105*(49), 19154–19159.

Beusen, B., Broekx, S., Denys, T., Beckx, C., Degraeuwe, B., Gijsbers, M., & Panis, L. I. (2009). Using on-board logging devices to study the longer-term impact of an eco-driving course. *Transportation Research Part D: Transport and Environment*, *14*(7), 514–520.

Bin, S., & Dowlatabadi, H. (2005). Consumer lifestyle approach to US energy use and the related CO_2 emissions. *Energy Policy*, *33*(2), 197–208.

Birol, F. (2017). CO_2 emissions from fuel combustion Highlights. Retrieved April 20, 2018, from https://www.iea.org/publications/freepublications/publication/CO2Emissions fromFuelCombustionHighlights2017.pdf.

Birrell, S. A., Young, M. S., & Weldon, A. M. (2013). Vibrotactile pedals: Provision of haptic feedback to support economical driving. *Ergonomics*, *56*(2), 282–292.

Birrell, S. A., Young, M. S., Jenkins, D. P., & Stanton, N. A. (2012). Cognitive work analysis for safe and efficient driving. *Theoretical Issues in Ergonomics Science*, *13*(4), 430–449.

Boden, T. A., Marland, G., & Andres, R. J. (2009). Global, regional, and national fossil-fuel CO_2 emissions. Carbon dioxide information analysis center, Oak ridge national laboratory, US department of energy, Oak Ridge, Tenn., USA.

Boriboonsomsin, K., Vu, A., & Barth, M. (2010). *Eco-driving: Pilot evaluation of driving behavior changes among us drivers*. University of California Transportation Center, Berkeley, CA.

Chan, C. C. (2007). The state of the art of electric, hybrid, and fuel cell vehicles. *Proceedings of the IEEE*, *95*(4), 704–718.

Crawford, J., & Neal, A. (2006). A review of the perceptual and cognitive issues associated with the use of head-up displays in commercial aviation. *The International Journal of Aviation Psychology*, *16*(1), 1–19.

Davidsson, S., & Alm, H. (2014). Context adaptable driver information–or, what do whom need and want when? *Applied Ergonomics*, *45*(4), 994–1002.

Degraeuwe, B., & Beusen, B. (2013). Corrigendum on the paper "Using on-board data logging devices to study the longer-term impact of an eco-driving course". *Transportation Research Part D: Transport and Environment*, *19*, 48–49.

Delhomme, P., Cristea, M., & Paran, F. (2013). Self-reported frequency and perceived difficulty of adopting eco-friendly driving behavior according to gender, age, and environmental concern. *Transportation Research Part D: Transport and Environment*, *20*, 55–58.

Dietz, T., Gardner, G. T., Gilligan, J., Stern, P. C., & Vandenbergh, M. P. (2009). Household actions can provide a behavioral wedge to rapidly reduce US carbon emissions. *Proceedings of the National Academy of Sciences*, *106*(44), 18452–18456.

Dogan, E., Bolderdijk, J. W., & Steg, L. (2014). Making small numbers count: Environmental and financial feedback in promoting eco-driving behaviours. *Journal of Consumer Policy*, *37*(3), 413–422.

Dula, C. S., & Geller, E. S. (2003). Risky, aggressive, or emotional driving: Addressing the need for consistent communication in research. *Journal of Safety Research, 34*(5), 559–566.

Franke, T., Arend, M. G., McIlroy, R. C., & Stanton, N. A. (2016). Ecodriving in hybrid electric vehicles–Exploring challenges for user-energy interaction. *Applied Ergonomics, 55*, 33–45.

Froehlich, J., Dillahunt, T., Klasnja, P., Mankoff, J., Consolvo, S., Harrison, B., & Landay, J. A. (2009, April). UbiGreen: Investigating a mobile tool for tracking and supporting green transportation habits. In *Proceedings of the SIGCHI Conference on Human Factors in Computing Systems.* (pp. 1043–1052). ACM.

Fuglestvedt, J. S., Shine, K. P., Berntsen, T., Cook, J., Lee, D. S., Stenke, A., & Waitz, I. A. (2010). Transport impacts on atmosphere and climate: Metrics. *Atmospheric Environment, 44*(37), 4648–4677.

Gregg, J. S., Andres, R. J., & Marland, G. (2008). China: Emissions pattern of the world leader in CO_2 emissions from fossil fuel consumption and cement production. *Geophysical Research Letters, 35*, L08806, doi:10.1029/2007GL032887

Heyes, D., Daun, T. J., Zimmermann, A., & Lienkamp, M. (2015). The virtual driving coach-design and preliminary testing of a predictive eco-driving assistance system for heavy-duty vehicles. *European Transport Research Review, 7*(3), 1–13.

Heyman, J., & Ariely, D. (2004). Effort for payment a tale of two markets. *Psychological Science, 15*(11), 787–793.

Hill, N., Brannigan, C., Smokers, R., Schroten, A., Van Essen, H., & Skinner, I. (2012). EU Transport GHG: Routes to 2050 ii. *final project report funded by the European Commission's Directorate-General Climate Action, Brussels.*

Hooker, J. N. (1988). Optimal driving for single-vehicle fuel economy. *Transportation Research Part A: Policy and Practice, 22*(3), 183–201.

Husnjak, S., Forenbacher, I., & Bucak, T. (2015). Evaluation of eco-driving using smart mobile devices. *PROMET-Traffic & Transportation, 27*(4), 335–344.

Jamson, A. H., Hibberd, D. L., & Merat, N. (2013). The design of haptic gas pedal feedback to support eco-driving. In *Proceedings of the Seventh International Driving Symposium on Human Factors in Driver Assessment, Training, and Vehicle Design.* (pp. 264–270). University of Iowa.

Kircher, K., Fors, C., & Ahlstrom, C. (2014). Continuous versus intermittent presentation of visual eco-driving advice. *Transportation Research Part F: Traffic Psychology and Behaviour, 24*, 27–38.

Kurani, K., Sanguinetti, A., & Park, H. (2015). "Actual Results May Vary": A Behavioral Review of Eco--Driving for Policy Makers. *White Paper, National Center for Sustainable Transportation, University of California at Davis, Davis, CA, July.*

Lai, W. T. (2015). The effects of eco-driving motivation, knowledge and reward intervention on fuel efficiency. *Transportation Research Part D: Transport and Environment, 34*, 155–160.

Larsson, H., & Ericsson, E. (2009). The effects of an acceleration advisory tool in vehicles for reduced fuel consumption and emissions. *Transportation Research Part D: Transport and Environment, 14*(2), 141–146.

Lauper, E., Moser, S., Fischer, M., Matthies, E., & Kaufmann-Hayoz, R. (2015). Psychological predictors of eco-driving: A longitudinal study. *Transportation Research Part F: Traffic Psychology and Behaviour, 33*, 27–37.

Lorf, C., Martínez-Botas, R. F., Howey, D. A., Lytton, L., & Cussons, B. (2013). Comparative analysis of the energy consumption and CO_2 emissions of 40 electric, plug-in hybrid electric, hybrid electric and internal combustion engine vehicles. *Transportation Research Part D: Transport and Environment, 23*, 12–19.

Madden, T. J., Hewett, K., & Roth, M. S. (2000). Managing images in different cultures: A cross-national study of color meanings and preferences. *Journal of International Marketing, 8*(4), 90–107.

Martin, E., Chan, N., & Shaheen, S. (2012). How public education on eco-driving can reduce both fuel use and greenhouse gas emissions. *Transportation Research Record: Journal of the Transportation Research Board, 2287*, 163–173.

McIlroy, R. C., & Stanton, N. A. (2017). What do people know about eco-driving? *Ergonomics, 60*(6), 754–769.

McIlroy, R. C., Stanton, N. A., & Harvey, C. (2014). Getting drivers to do the right thing: A review of the potential for safely reducing energy consumption through design. *Intelligent Transport Systems, IET, 8*(4), 388–397.

McIlroy, R. C., Stanton, N. A., Godwin, L., & Wood, A. P. (2016). Encouraging eco-driving with visual, auditory and vibrotactile stimuli. *IEEE Transactions on Human Machine Systems, 47(5)*, 661–672.

Mensing, F., Bideaux, E., Trigui, R., Ribet, J., & Jeanneret, B. (2014). Eco-driving: An economic or ecologic driving style? *Transportation Research Part C: Emerging Technologies, 38*, 110–121.

Meschtscherjakov, A., Wilfinger, D., Scherndl, T., & Tscheligi, M. (2009, September). Acceptance of future persuasive in-car interfaces towards a more economic driving behaviour. In *Proceedings of the 1st International Conference on Automotive User Interfaces and Interactive Vehicular Applications.* (pp. 81–88). ACM.

Netherlands Environmental Assessment Agency. (2017) CO_2 time series 1990–2015 per region/country. Retrieved from http://edgar.jrc.ec.europa.eu/overview.php?v=CO2ts1990-2015 on 04/04/2018.

Rasmussen, J. (1986). *Information processing and human-machine interaction: An approach to cognitive engineering.* New York: North-Holland.

Recarte, M. A., & Nunes, L. M. (2003). Mental workload while driving: Effects on visual search, discrimination, and decision making. *Journal of Experimental Psychology: Applied, 9*(2), 119.

Rolim, C., Baptista, P., Duarte, G., Farias, T., & Pereira, J. (2016). Impacts of delayed feedback on eco-driving behavior and resulting environmental performance changes. *Transportation Research Part F: Traffic Psychology and Behaviour, 43*, 366–378.

Rouzikhah, H., King, M., & Rakotonirainy, A. (2013). Examining the effects of an eco-driving message on driver distraction. *Accident Analysis & Prevention, 50*, 975–983.

Saboohi, Y., & Farzaneh, H. (2009). Model for developing an eco-driving strategy of a passenger vehicle based on the least fuel consumption. *Applied Energy, 86*(10), 1925–1932.

Shabbir, W., & Evangelou, S. A. (2014). Real-time control strategy to maximize hybrid electric vehicle powertrain efficiency. *Applied Energy, 135*, 512–522.

Sivak, M., & Schoettle, B. (2012). Eco-driving: Strategic, tactical, and operational decisions of the driver that influence vehicle fuel economy. *Transport Policy, 22*, 96–99.

Skeie, R. B., Fuglestvedt, J., Berntsen, T., Lund, M. T., Myhre, G., & Rypdal, K. (2009). Global temperature change from the transport sectors: Historical development and future scenarios. *Atmospheric Environment, 43*(39), 6260–6270.

Staubach, M., Schebitz, N., Fricke, N., Schießl, C., Brockmann, M., & Kuck, D. (2014). Information modalities and timing of ecological driving support advices. *IET Intelligent Transport Systems, 8*(6), 534–542.

Stillwater, T., & Kurani, K. S. (2013). Drivers discuss eco-driving feedback: Goal setting, framing, and anchoring motivate new behaviors. *Transportation Research Part F: Traffic Psychology and Behaviour, 19*, 85–96.

Strömberg, H., Karlsson, I. M., & Rexfelt, O. (2015). Eco-driving: Drivers' understanding of the concept and implications for future interventions. *Transport Policy, 39*, 48–54.

Summala, H., Lamble, D., & Laakso, M. (1998). Driving experience and perception of the lead car's braking when looking at in-car targets. *Accident Analysis & Prevention, 30*(4), 401–407.

Thornton, J., & Covington, H. (2015). Climate change before the court. *Nature Geoscience, 9*(1), 3–5.

Tulusan, J., Staake, T., & Fleisch, E. (2012, September). Providing eco-driving feedback to corporate car drivers: What impact does a smartphone application have on their fuel efficiency? In *Proceedings of the 2012 ACM Conference on Ubiquitous Computing.* (pp. 212–215). ACM.

Van der Voort, M., Dougherty, M. S., & Van Maarseveen, M. (2001). A prototype fuel-efficiency support tool. *Transportation Research Part C: Emerging Technologies, 9*(4), 279–296.

Van Erp, J. B. F., & Padmos, P. (2003). Van image parameters for driving with indirect viewing systems. *Ergonomics, 46*(15), 1471–1499.

Vandenbergh, M. P., & Steinemann, A. C. (2007). The carbon-neutral individual. *NYUL Rev, 82*, 1673.

Vandenbergh, M. P., Barkenbus, J., & Gilligan, J. (2007). Individual carbon emissions: The low-hanging fruit. *UCLA L. Rev, 55*, 1701.

Wåhlberg, A. E. (2007). Long-term effects of training in economical driving: Fuel consumption, accidents, driver acceleration behavior and technical feedback. *International Journal of Industrial Ergonomics, 37*(4), 333–343.

Wickens, C. D. (2002). Multiple resources and performance prediction. *Theoretical Issues in Ergonomics Science, 3*(2), 159–177.

Wickens, C. D. (2008). Multiple resources and mental workload. *Human Factors: The Journal of the Human Factors and Ergonomics Society, 50*(3), 449–455.

Windecker, A., & Ruder, A. (2013). Fuel economy, cost, and greenhouse gas results for alternative fuel vehicles in 2011. *Transportation Research Part D: Transport and Environment, 23*, 34–40.

Young, K. & Regan, M. (2007). Driver distraction: A review of the literature. In: I.J. Faulks, M. Regan, M. Stevenson, J. Brown, A. Porter & J.D. Irwin (Eds.). Distracted Driving. Sydney, New South Wales: Australasian College of Road Safety. 379–405.

Young, M. S., Birrell, S. A., & Stanton, N. A. (2009). Design for smart driving: A tale of two interfaces. In *Engineering psychology and cognitive ergonomics.* 477–485. Springer Berlin Heidelberg, Germany.

Zarkadoula, M., Zoidis, G., & Tritopoulou, E. (2007). Training urban bus drivers to promote smart driving: A note on a Greek eco-driving pilot program. *Transportation Research Part D: Transport and Environment, 12*(6), 449–451.

Zhao, X., Wu, Y., Rong, J., & Zhang, Y. (2015). Development of a driving simulator based eco-driving support system. *Transportation Research Part C: Emerging Technologies, 58*, 631–641.

2 Applying Cognitive Work Analysis to Understand Fuel-Efficient Driving

INTRODUCTION

Anthropometric climate change is, and will continue to be, a defining truth of the 21st century (Thornton & Covington, 2015). Greenhouse gases (GHGs) for example carbon dioxide (CO_2) and nitrous oxides (NOx) generated by human action, for example transport and manufacturing, play a key role in influencing global temperature and weather conditions. Whilst it can be difficult for individuals to see the impact of their own behaviour, approximately 30–40% of total GHG emissions can be directly attributed to individual energy needs supporting 21st-century western lifestyles (Vandenbergh et al., 2007). Overall emissions can however be significantly reduced if individuals modify their behaviour to act in a more environmentally friendly manner (Dietz et al., 2009).

Transport is a prolific source of pollution, especially GHG emissions. In the European Union (EU), the quantity of GHG emitted directly as a consequence of transport comes second only to electricity generation (Hill et al., 2012). Road transportation, typically personal cars, is the highest contributor to this statistic, accounting for approximately 75% of all transport GHG emissions (Hill et al., 2012). Similar patterns of emissions can be seen in the United States of America, whereby car use accounts for between 32% and 41% of the US's total CO_2 emissions (Bin & Dowlatabadi, 2005; Vandenbergh & Steinemann, 2007). Barkenbus (2010) estimates that approximately 8% of the worlds' current CO_2 emissions are a result of transportation. Minimising transportation, and specifically, automobile-related emissions should therefore be considered a low-hanging fruit to reduce current levels of GHG emissions. Due to the vast number of road vehicles currently active, with the UK alone having an estimated 30.3 million private cars on the road (Department of Transport Vehicle Licensing Statistics: Quarter 4 (Oct–Dec) 2015), small-scale savings made within this sector can have a significant impact.

Whilst limiting the number of car journeys would be the most dramatic way to reduce automotive vehicle-related GHG emissions (Strömberg et al., 2015), this approach is not viable. Two alternate approaches can however enable the use of cars whilst reducing fuel use and emissions. The first approach is the development and implementation of new in-vehicle technology and fuel-efficient drivetrains than those currently available. Improvements in vehicle drivetrain efficiency can be seen when considering Hybrid and Plug-in Hybrid Electric Vehicles (PHEVs), such as

DOI: 10.1201/9781003081173-2

the Toyota Prius, Nissan Leaf, and Mitsubishi PHEV. These vehicles use developments in technology to reduce emissions and achieve greater fuel efficiency, with Mitsubishi PHEV claiming to achieve 157 miles per gallon, considerably greater than similar internal combustion engine (ICE) vehicles. The second approach is modifying driver behaviours, in order to encourage the adoption of fuel-efficient, eco-driving techniques. Barkenbus (2010) proposed that eco-driving is characterised by behaviours such as modest acceleration, early gear changes, limiting the engine to approximately 2,500 revolutions per minute (RPM), anticipating traffic flow to minimise braking, driving below the speed limit, and limiting unnecessary idling. A key advantage of encouraging eco-driving behaviours is that these techniques can be employed by all drivers, and are not reliant on significant financial investment, such as would be necessary to purchase a new vehicle, with a more fuel-efficient drivetrain. Consequently, encouraging eco-driving is an appropriate technique for owners of older cars and those with limited disposable income. Although eco-driving training has been demonstrated to offer significant fuel saving benefits (Wu et al., 2017), it is also clear that many drivers do not engage in such behaviours despite previous awareness of eco-driving techniques (Pampel et al., 2017). Previous work (Allison & Stanton, 2019 McIlroy et al., 2013) has highlighted that eco-driving should be supported and encouraged regularly via the use of in-vehicle interfaces to maintain long-term effectiveness. Identifying the constraints that operate around car-use and driving behaviours, can enable the informed development of low-cost interfaces supporting the maintenance of such behaviours.

In-car interfaces can be implemented to facilitate drivers' awareness of their current behaviours (Oinas-Kukkonen & Harjumaa, 2009) as well as support and train drivers to engage in more environmentally friendly driving behaviours (Tulusan et al., 2012). The adoption of eco-driving behaviours can lead to a reduction in fuel use and subsequently a reduction of CO_2 and NOx emissions. Such interfaces have received positive reviews, for their potential in reducing GHG emissions (Carsten & Tate, 2005), but concerns relating to their use have arisen due to the possible negative implications they can have on safety due to potential for such displays to distract drivers (Cacciabue & Saad, 2008; Young et al., 2011). Due to this dichotomy, a focus on the design of potential interfaces is considered a priority for research in a variety of transport interface research fields, commonly following a users' centred design approach (Cacciabue & Martinetto, 2006; Fénix et al., 2008).

COGNITIVE WORK ANALYSIS

Cognitive work analysis (CWA) is a framework for understanding complex sociotechnical systems, characterised by close interactions of people and technology (Stanton & Bessell, 2014). Originally developed for use in the nuclear power industry (Rasmussen, 1986), it has been suggested that CWA can act as a key tool when developing and designing novel systems (Rasmussen et al., 1990). CWA seeks to understand the constraints that frame a working system, understanding what is required of the system as well as what is both possible and not possible within the confines of the system (Kant, 2017). By focussing on the constraints, the analysis seeks to understand and support workers needs for improved efficiency and safety (Stanton et al., 2013). CWA has been used for understanding a variety of complex systems including military planning systems (Jenkins et al., 2008; Stanton & McIlroy, 2012), team design (Naikar et al., 2003), the aviation domain (Stanton et al., 2016), and even the Apple iPod personal music device (Cornelissen et al., 2013). Drawing upon foundations in ecological psychology, general systems thinking, and adaptive control systems (Fidel & Pejtersen, 2004), CWA has developed into a domain agnostic and highly flexible method that can be utilised to understand both current work domains and explore the potential of future developments. Naikar and Lintern (2002) suggest that CWA is an ideal method for envisioning revolutionary design as it allows a focus on the fundamental requirements of the system. CWA is an appropriate method for the challenge of supporting eco-driving behaviours as it can be used for both systems design of and inform the design of future human-machine interfaces (Van Westrenen, 2011). Indeed, previous work has used independent components of the CWA process to support eco-driving including developing an Abstraction Hierarchy (Birrell et al., 2012) and Decision Ladders (McIlroy & Stanton, 2015a).

The complete CWA process comprises of five key phases, Work Domain Analysis (WDA), Control Task Analysis (ConTA), Strategies Analysis (StrA), Social Organisation and Cooperation Analysis (SOCA), and Worker Competencies Analysis (WCA) (Vicente, 1999; McIlroy & Stanton, 2011). The primary focus of the WDA is the development of an abstraction hierarchy. The abstraction hierarchy aims to map the proposed system on multiple conceptual levels, ranging from its reason for existing to the physical objects that comprise the system (Naikar, 2013). ConTA examines the constraints that operate on a system within set conditions. ConTA can be seen as defining the available inputs and end goal of a system (Naikar, 2006). The common tool for ConTA is that of the decision ladder (Vicente, 1999), and examines the steps, and potential shortcuts, a user follows when operating a system. Phase 3 is the StrA component. It has been argued (Naikar, 2006) that the strategies individuals use to operate a system can vary significantly under different operating conditions, for example as a result of time pressure or during high-stress situations, such as a safety-critical event. StrA often takes the form of flow maps, defining start and end states of the system, and describing the potential tasks required to transform the system from the start to the desired end state (Ahlstrom, 2005; Naikar, 2006). The fourth phase of CWA, SOCA, examines the allocation of tasks within the system, accounting for both human and technological agents on an equal level (Vicente, 1999). SOCA maps the responsibility of given tasks to different actors and acts as a way to reallocate tasks to improve efficiency and safety. The dynamic allocation of function can be applied to all elements of the CWA process, including the abstraction hierarchy, Contextual Activity Template (CAT), StrA, and the WCA. The final phase of the CWA framework is the WCA. Kilgore and St-Cyr (2006) suggest that this stage of CWA focusses on psychological constraints of the users of systems. WCA often draws upon the Skills, Rules, and Knowledge taxonomy (SRK) (Vicente, 1999) to map out the level of cognitive effort applied to the different tasks.

Despite the final outcome of the CWA analysis not being a complete workable design of the envisaged system or interface, it is argued that mapping the constraints and requirements can produce a final system more suitable to end-user requirements, reducing the need for future iterative design stages (McIlroy & Stanton, 2011). CWA offers analysts a technology-agnostic approach to consider a system, allowing for the consideration of both technology and human agents in the same analysis (Vicente, 1999; Jenkins et al., 2008). This makes it an ideal approach for the development of novel technology as well as a tool to consider the constraints for a new addition or interface within a previously established domain. Driving is one such domain whereby drivers must interact with in-built vehicle mechanical systems, other road users and, increasingly, in-vehicle technology, such as driver-assist technology. This chapter will document the development of a CWA to support fuel-efficient driving and lay down the design constraints necessary to develop interfaces that support fuel-efficient driving.

METHOD

Whilst a complete CWA can be seen as an extensive and time consuming task, requiring numerous iteration and consultation, individual elements of the CWA process can be completed independent of the full analysis and subsequently refined and validated (McIlroy & Stanton, 2011). Lintern et al. (2004) argued that the WDA component of CWA provides an ideal starting point. As the key outcome of the WDA is the abstraction hierarchy, initial focus was given to the development of this metric. Although previous research (Birrell et al., 2012) has developed an abstraction hierarchy to support eco-driving, considerable technical developments have occurred within the automotive domain since this work, including considerable progress relating to the development and large-scale use of hybrid vehicles. It was therefore deemed prudent to reconsider this work in light of these technical developments.

The abstraction hierarchy was created over the course of a daylong workshop with a series of academics with a research interest in automotive fuel efficiency. Research participants comprised of nine academic staff from the University of Southampton, University College London, and Imperial College London. Participants were aged between 27–57 years old ($M = 36.3$, $S.D. = 10.8$), and all held doctoral degree-level qualifications within their respective fields of expertise. The workshop drew together individuals with an extensive knowledge of vehicle dynamics, control engineering, drivetrain optimisation, traffic operations, eco-driving, and fuel usage. The attendees held a mixture of backgrounds, including Engineering, Human Factors, and Traffic Modelling. This enabled a consideration of factors that would be unknown to any single member of the team. As Sharp and Helmicki (1998) suggest, CWA is heavily dependent on the skills and knowledge of the analysts and the resources at their disposal. Analysts' understanding and knowledge of the system functioning are therefore key in developing a useful and complete CWA. To facilitate the workshop, two members of the team had extensive knowledge and experience of the CWA process.

Following the initial workshop, the abstraction hierarchy was formally constructed based upon ideas discussed during the workshop and any remaining links between different elements were generated by the Human Factors team. The initial workshop team subsequently validated the generated links and elements during a follow-up workshop. For the validation exercise, each team member was presented with the completed abstraction hierarchy, including the completed means-end analysis, and were asked to exhaustively check each object and link with the hierarchy. To facilitate this process, each workshop participant was presented with a matrix of each level pairing from the abstraction hierarchy, for example, "Functional Purpose" paired with "Values and Priorities" and asked to mark within the matrix whether each of the current connections were valid or whether a new connection was needed. Following this process, each modification, i.e. removal or addition of connections, was discussed in depth with the wider team and group consensus was reached for each paring.

Once the agreed upon abstraction hierarchy was validated, the remaining CWA phases were completed independently by the Human Factors team and subsequently validated by the larger workshop team upon the completion of each stage.

RESULTS AND DISCUSSION

Work Domain Analysis

When constructing an abstraction hierarchy for fuel-efficient driving, the primary identified functional purposes of the system were "Save Energy" and "Reduce Emissions (CO_2 and NOx)". A third potential functional purpose was added, "Getting From A to B" however questions were raised by members of the team whether this was a function of the proposed device or a by-product of being a device designed to operate within a car. Although consensus was reached, with the majority of the workshop team, it was a considerable point of discussion. The group continued to discuss the Values and Priorities layer, the metrics for identifying whether develop device would be successful at its previously defined Functional Purposes. Eight Values and Priorities were identified, "Optimise Vehicle Range"; "Reduce Fuel Usage"; "Minimise Traffic Delay"; "Minimise Congestion"; "Optimise Driver Satisfaction"; "Optimise Travel Time"; "Reduce NOx"; and "Reduce CO_2".

Rather than continuing the top-down approach, which had driven the discussion up to this point, the team switched to a bottom up strategy to discuss the physical objects that would be required by the potential system and elements which the system could interact with. This primarily included sensors which the vehicle would need in order to be aware of its surrounding environment, power systems and drivetrain to propel the vehicle, direct vehicular controls, and ancillary systems which require energy in order to operate, but are not part of the fundamental requirements for vehicle motion, for example air condition and heating.

Once the Physical Objects within the system had been defined, the team progressed to discuss the Object Related Properties layer within the Hierarchy. This layer focusses primarily on the function of the physical objects of the system, for example "Air Conditioning", a physical object, influences "Cabin Temperature" and controls "Cabin Humidity". The development of the Object Related Properties layer was also key in identifying Physical Objects that had not been initially considered, prompting further discussion.

The final stage to be completed was the Purpose Related Functions. This layer seeks to bridge the objective elements of the hierarchy with the larger gestalt aims of the system under investigation. The functions identified related to identifying "Road Attributes", "Objects in the World", "Control Vehicle Motion", and ensuring "Driver Comfort" and "Passenger Comfort".

Whilst the complete mapped abstraction hierarchy is too large to be presented within this volume, the complete list of items that abstraction hierarchy is comprised of is presented within Table 2.1. In addition, the purpose of each Physical Object, and its corresponding impact on fuel usage is presented in Table 2.2, alongside a corresponding reference. To test the suitability and completeness of the identified abstraction hierarchy, an exhaustive means ends analysis was completed following the why-what-how triad approach (Rasmussen et al., 1994; Vicente, 1999). Within the abstraction hierarchy, it possible to nominate any item within the matrix and ask the question, *"what does this do?"* When considering all connections in the layer immediately above the node, it is possible to answer the question, *"why does it do this?"* When considering all connections in the layer immediately below, it is possible to answer the question, *"how does it achieve this?"* This ensured that all required connections were adequately captured and addressed. Following this analysis, the final abstraction hierarchy was produced. Due to the number of items and connections that were identified within the abstraction hierarchy, Figure 2.1 presents a subsection of the abstraction hierarchy exploring the connections and affordances of the physical object "GPS" to illustrate the connections which exist.

TABLE 2.1
Object Related Processes and Physical Objects Identified within the Abstraction Hierarchy

Functional Purposes	• Save Energy
	• Reduce Emissions (CO_2 and NOx)
	• Getting from A to B
Values and Priorities	• Optimise Vehicle Range
	• Reduce Fuel Usage
	• Minimise Traffic Delay
	• Minimise Congestion
	• Optimise Driver Satisfaction
	• Optimise Travel Time
	• Reduce NOx
	• Reduce CO_2

(*Continued*)

TABLE 2.1 (Continued)
Object Related Processes and Physical Objects Identified within the Abstraction Hierarchy

Purpose Related Function	• Detect and Present Road Attributes • Detect and Present Objects in the World • Control Vehicle Motion • Provide Motive Force • Support Driver Comfort • Support Passenger Comfort
Object Related Processes	• Display Vehicle Speed • Display Vehicle RPM • Control Cabin Humidity • Alert of Hazards • Detect Traffic Jams • Provide Information on Other Vehicle Behaviours • Hold Knowledge of Most Fuel-Efficient Path • Understand Own Vehicle Position • Understand Own Vehicle Motion • Detect Other Road Users • Hold Knowledge of Weather Condition & Forecast • Hold Knowledge of Shortest Path • Hold Knowledge of Fastest Path • Detect Pedestrians • Control Lighting • Hold Knowledge of Road Gradient • Hold Knowledge of Road Width • Detect Infrastructure • Present Speed Limit • Present Information on Traffic Lights • Control Acceleration • Control Vehicle Speed • Control Vehicle Lane Position • Control Vehicle Heading • Control Vehicle Headway • Present Vehicle Path • Provide a Path • Provide Energy • Control Temperature of Cabin • Provide Information on Range • Predict Closure of Roadway • Detect Speed of Other Vehicles • Smooth Motion • Detect Distance of Other Vehicles • Maintain Road Adherence • Provide Information on Intersections • Provide Information on Road Curvature • Present Battery (State of Charge) • Present Battery (State of Health)

(Continued)

TABLE 2.1 (Continued)
Object Related Processes and Physical Objects Identified within the Abstraction Hierarchy

Physical Objects	
	- Hybrid Electric Vehicles
	- Torque Converter
	- Heating
	- Clutch
	- Energy Recovery (Brake)
	- Battery
	- Vehicle Lights
	- Start/ Stop
	- V2X Communication
	- Internal Combustion Engine
	- Fuel
	- Automation
	- Infrastructure (Road)
	- GPS
	- Camera/Vision Systems
	- Current Weather
	- Lidar
	- Road Surface
	- Road Markings
	- Meteorological Service
	- Road Network Map
	- Radar
	- Gearbox
	- Brake
	- Accelerator
	- Instrument Cluster
	- Steering Wheel
	- Air Conditioning
	- Tyres

TABLE 2.2
Description of the Physical Objects Layer, and Their Influence on Fuel Usage

Physical Object	Description	Impact on Fuel Economy	Reference
Hybrid Electric Vehicles	Vehicle with multiple power sources to power the drivetrain uses both fuel and stored electric energy.	Electrical energy is not fuel, allowing engine to run more efficiently as the battery makes up any energy deficit.	Ehsani et al. (2018)

(Continued)

TABLE 2.2 (Continued)
Description of the Physical Objects Layer, and Their Influence on Fuel Usage

Physical Object	Description	Impact on Fuel Economy	Reference
Torque Converter	Present within automatic transmission vehicles instead of a clutch, power from the engine transferred to the gearbox via torque converter.	More efficient torque converter leads to increased power to the gearbox resulting in reduced fuel usage.	Robinette et al. (2011)
Heating	Provides passenger comfort.	Within Hybrid and electrical vehicles typically powered by the electrical system requiring energy. Within internal combustion engines typically powered by waste heat from the engine.	Horrein et al. (2016)
Clutch	Within manual transmission vehicles. Controls power flow between engine and gearbox.	Control of clutch affects fuel usage by influencing RPM and enabling coasting behaviours.	Barkenbus (2010)
Energy Recovery (Brake)	Available within electric and hybrid vehicles, vehicle is capable of acting as a generator to recharge battery.	Fuel is not required to solely power the vehicle, allowing reduced fuel usage.	Ehsani et al. (2018)
Battery	Stores electrical energy to allow electrical systems to function.	Provides an alternative energy source for the vehicle.	Ehsani et al. (2018)
Vehicle Lights	Provides visibility to the driver and provide warning to other drivers about driver intentions.	Requires electrical energy to function.	Gillespie (1992)
Start/ Stop	Turns engine off when vehicle is not in motion.	Stops fuel being used due to engine idling.	Barkenbus (2010)
V2x Communication	Provides driver with information about upcoming road events and other vehicles.	Opportunity to guide driver actions, for example, use of alternative route to avoid congestion or slowing the vehicle prior to a required stop due to traffic lights.	LaClair et al. (2014)

(Continued)

TABLE 2.2 (Continued)
Description of the Physical Objects Layer, and Their Influence on Fuel Usage

Physical Object	Description	Impact on Fuel Economy	Reference
Internal Combustion Engine	Uses fuel to provide motive force to the vehicle.	Overall engine efficiency affects fuel usage.	Gillespie (1992)
Fuel	Provides chemical energy for conversion into electrical and kinetic energy to power vehicle.	Is fuel.	Gillespie (1992)
Automation	Option to allow the vehicle to perform tasks without direct human action/involvement.	Provides potential for more efficient journey parameters including optimisation which drivers would otherwise struggle with.	Miller and Heard (2016)
Infrastructure (Road)	Road objects and elements that control the flow of traffic.	Applies limitations to potential actions of the driver, which can act to both improve fuel efficiency (e.g. improving flow) or penalise fuel efficiency (traffic lights).	LaClair et al. (2014)
GPS	Possess a map for the road environment and provide position and time information to guide route decisions.	Can generate most fuel-efficient route information and effect subsequent route decisions.	Tunnell et al. (2018).
Camera/ Vision System	Provide information on other objects within the road environment.	Provide advanced warning of potential hazards allowing for smoother driving style.	Tunnell et al. (2018).
Current Weather	Physical weather conditions which would affect both road surface and visibility.	Potential impact on fuel efficiency due to changes in road rolling resistance and potential impacts on potential speed choices.	LaClair et al. (2014)
Lidar	Sensor to measure distance to surrounding objects and vehicles.	Potential for use in adaptive cruise control systems. Provide guidance on potential behaviours.	Tunnell et al. (2018).
Road Surface	Provide low rolling resistance substrate to allow vehicle to drive along.	Different surfaces have different rolling resistance affecting overall fuel usage.	Gillespie (1992)

(Continued)

TABLE 2.2 (Continued)
Description of the Physical Objects Layer, and Their Influence on Fuel Usage

Physical Object	Description	Impact on Fuel Economy	Reference
Road Marking	Denote route information to allow multiple vehicles to cooperate within a limited space.	Provide limitation on driver actions and control potential driver behaviour.	Takezaki et al. (2000)
Meteorology Service	Provide information on upcoming weather. Provide warning about road condition.	Potential to affect timing of potential drive and subsequent driver behaviour.	LaClair et al. (2014)
Road Network Map	Provide information on road network. Potential to provide information on traffic levels.	Potential to affect route decisions.	LaClair et al. (2014)
Radar	Sensor which measures velocity of and distance to surrounding objects and vehicles.	Potential for use in adaptive cruise control systems. Provide guidance on potential behaviours.	Tunnell et al. (2018).
Gearbox	Selects different engine operating speeds to control overall RPM control torque to the driven wheels.	Allows selection of appropriate RPM ratings at different speeds.	Barkenbus (2010)
Brake	Provide ability to decrease vehicle speed.	Kinetic energy is converted into heat which is dissipated and wasted.	Barkenbus (2010)
Accelerator	Provide ability to increase vehicle speed.	Requires fuel to be converted into kinetic energy.	Barkenbus (2010)
Instrument Cluster	Provide information to the driver including speed and RPM.	Option to influence driver actions and provide feedback to the driver.	Jamson et al. (2015)
Steering Wheel	Allow driver to control direction of travel.	Provide options to allow driver to change route and avoid obstacles.	Gillespie (1992)
Air Conditioning	Converts engine power for internal cabin climate control.	Requires energy, generated from fuel, to run, varying efficiency impacts overall fuel usage.	Yan et al. (2018)
Tyres	Increases adhesion of the car to the road by providing high friction between wheel and road surface.	More rigid/inflated tyres provide greater fuel economy as subsequently less rolling resistance.	Gillespie (1992)

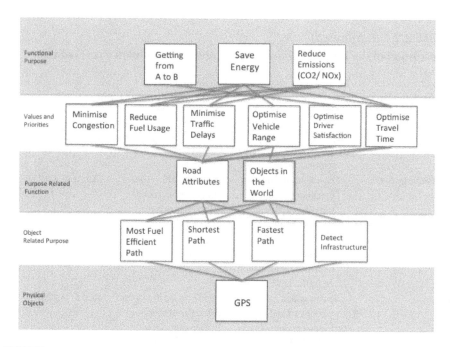

FIGURE 2.1 Excerpt from the completed Abstraction Hierarchy.

Taking the physical object of in-vehicle global positioning system (GPS), this object can be used to afford knowledge of the most "fuel-efficient path" to a given destination, the "shortest path", in terms of distance, to a given destination, and the "fastest path" to a given destination, in terms of travel time. The GPS system also affords the ability to "detect infrastructure", including traffic lights, junction, roundabouts, and speed limits, as each of these factors will influence the potential paths the vehicle can take. Knowledge of vehicle paths, be it most fuel-efficient, shortest, or fastest, are linked to "road attributes" and constrained by "objects in the world". The ability to detect infrastructure is linked to knowledge of "objects in the world". If a system has knowledge of road attributes, and objects in the world, which could, potentially include other road users, it will be possible to generate a route that "minimises congestion", "minimises fuel use", "minimises traffic delays", "optimises vehicle range", "optimises driver satisfaction", and "optimises traffic time". Because of these advantages, the system assists the driver in completing their journey and "get from A to B", but also "save energy" and "reduce emissions (CO_2 and NOx)". Similar thought processes can be applied to every object within the abstraction hierarchy to generate the complete set of links through each of the physical objects to the overall functional purposes of the system.

CONTROL TASK ANALYSIS

The ConTA builds on the ideas presented within the WDA, to consider temporal constraints on a system, that is, when key activities using the system would take place. A common tool for achieving this goal is the CAT (Naikar et al., 2006). The key objective of the CAT is to identify whether a task is typically completed within a given situation, whether a task can be completed within the situation, or whether the task is not possible within the situation. Within this framework, the work functions of a system are compared across considered across multiple potential situations, which may be as a consequence of spatial or temporal constraints (Rasmussen et al., 1994; Naikar et al., 2006). Defining appropriate work functions, however, can be a great challenge. Although a departure from traditional CWA approaches, Stanton and Bessell (2014) suggest that the Object Related Processes, identified within the abstraction hierarchy, are suitable for the definition of systems work functions. Whilst this may not hold true within large-scale systems with multiple supervisory operators, within the context of driving with a single operator this approach was deemed appropriate and has been adopted within the current analysis. To achieve this goal, a matrix was developed whereby each Object Related Process identified within the abstraction hierarchy was plotted against a variety of road situations that a driver could be faced with. The matrix was then overlaid with box and whisker plots to map whether the Object Related Process could be completed in the given situation. A box within a cell represents that the task can be completed within the situation; a circle or whisker indicates that an activity can be performed within the current situation and typically is and a cell with no markings indicates that an activity is not possible with the situation.

The development of the situations that would be considered was a key undertaking within this stage of the analysis. Situations that were initially considered were based on the generic road types that a driver could experience, including Motorway (Clear), Motorway (Congested), Urban, and Countryside. It quickly became apparent however that road type was largely not a key determinant factor to the identified Object Related Processes. Although minor differences were observed, for example traffic lights are not present within motorway driving, the majority of driving related tasks are independent of road type. A key example of this is the Object Related Process "Control vehicle heading", controlling vehicle heading is important for all scenarios where the vehicle is in motion, and not limited to a specific road type. To overcome this limitation, the holistic nature of a car journey was considered. The analysis was therefore widened to include specific journey scenarios, which could be encountered during a standard drive, including traveling on a motorway slip road and being on a roundabout. Figure 2.2 presents an excerpt of the completed CAT analysis demonstrating the different situations a driver could face. In total 23 different situations were considered, Motorway Clear; Motorway Congested; Urban; Major A-road; Country Road; Junction; Rural; Residential; Planning Journey; In-car Pre engine start; Waiting at Traffic Lights; Waiting at a Junction; On a Slip Road; Post Journey, Engine Turned off; Pre Journey, Engine Started, Handbrake on; On a Roundabout; Initial Acceleration from Stationary; Cruising/Steady Speed; Overtaking; Parking; Emergency Stopping; Reversing; and General Braking.

From Figure 2.2, it can be seen that for the tasks of "display vehicle speed", "display vehicle RPM", "Alert of Hazards", "Detecting Traffic Jams", "Providing information on Other Vehicle Behaviours", "Own Vehicle Position", "Own Vehicle motion", and "Other Road Users" are all typically applicable to the situations of driving on

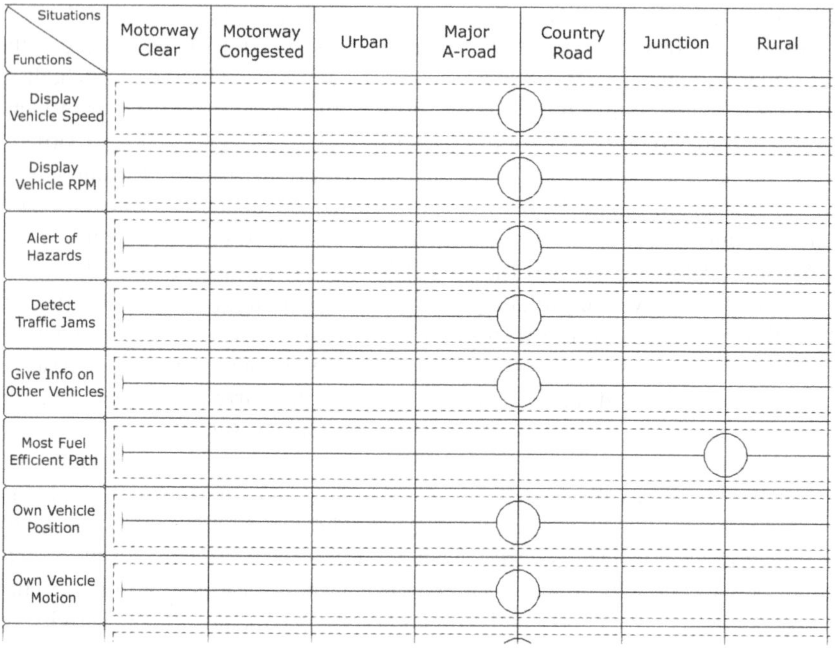

32 Assisted Eco-Driving

Situations / Functions	Motorway Clear	Motorway Congested	Urban	Major A-road	Country Road	Junction	Rural
Display Vehicle Speed					○		
Display Vehicle RPM					○		
Alert of Hazards					○		
Detect Traffic Jams					○		
Give Info on Other Vehicles					○		
Most Fuel Efficient Path						○	
Own Vehicle Position					○		
Own Vehicle Motion					○		

FIGURE 2.2 Excerpt from the completed CAT analysis.

"Motorway (Clear)", "Motorway (Congested)", "Urban", "Major A-Road", "Country Road", "Junction", "Rural", and "Residential", however are not experienced when a driver is initially planning their journey. The task of most fuel-efficient path however can occur in all of the aforementioned scenarios, including "Planning Journey". The task of "Traffic Lights" is also different as traffic lights do not appear in motorways, consequently cannot be encountered during this situation. Traffic lights also typically do not appear on major A-roads, although it is possible to encounter the traffic lights on this road type, this is not typical. Traffic lights also are not encountered when planning a journey.

The CAT analysis highlighted how important the majority of the identified Object Related Processes are to the general task of driving. This finding was not initially anticipated, and although it could be argued such finding makes the analysis superfluous, this finding is useful when considering the value of presenting situation tailored information. The analysis revealed how key vehicle metrics available to the driver are independent of immediate need. One example of this is "Display Vehicle Speed", which is available to the driver in all situations post engine start, including when the vehicle is stationary, hence has no recorded speed and speed information is unnecessary. With the greater use of LCD displays within vehicles, future development could be influenced by knowledge of current context, supported by vehicle information data, including speed and on-board GPS, and by the potential of V2X communication. The push for context-driven information has been supported within previous work exploring drivers' information desires (Davidsson & Alm, 2014).

Across a series of interviews with 33 drivers, Davidsson and Alm (2014) found that information on which drivers ranked as highly important in one context was not required in a different context. Taken together with the results of the current study, leveraging the role of context could play a central role in supporting fuel-efficient behaviours.

It is typical to progress the insights of the ConTA at this point to consider constraints in terms of decision making processes. This analysis is typically achieved using Decision Ladders (Vicente, 1999). The decision ladder presents a linear sequence of information processing steps, with novices following a linear process through all the steps and expert users able to make cognitive shortcuts through the steps. Whilst this stage would present a valuable addition to the current work, extensive work exploring the decision ladders approach within eco-driving has already been completed (McIlroy & Stanton, 2015a). Due to the depth of this analysis, there is little need within the current analysis to revisit this topic area within the current study. The decision ladders developed by McIlroy and Stanton (2015a) were therefore used within the current study when progressing through the remaining CWA stages.

STRATEGIES ANALYSIS

To progress through the StrA component of the CWA process, a series of simplified flow maps (Ahlstrom, 2005) were produced. Although the use of extensive flow diagrams has been recommended for this stage of the investigation (Vicente, 1999), the lack of clear guidance in how these should be produced and a lack of tools to facilitate the process hampers the adoption of this approach. Simplified flow maps are designed to present the strategies that individuals can use to achieve a set goal. Where appropriate, an individual may have access to different strategies at different times, dependent on external factors, for example, time constraints or accessibility of required tools. Simplified flow maps were created for the tasks of deceleration, acceleration, managing distance to a preceding vehicle (headway), and maintaining current speed. The developed flow maps were based upon the decision ladders presented by (McIlroy & Stanton, 2015a), and validated by the initial workshop team upon completion. The simplified flow maps are presented below for the scenarios of deceleration to a lower speed (Figure 2.3.1); acceleration to a higher speed (Figure 2.3.2); managing headway (Figure 2.3.3); and maintaining current speed (Figure 2.3.4).

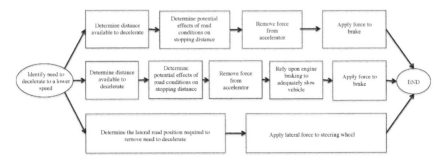

FIGURE 2.3.1 Simplified Flow map for the task of deceleration to a lower speed.

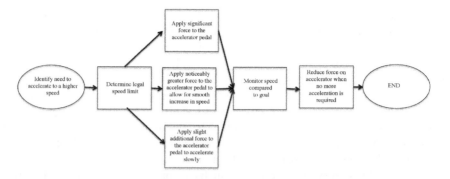

FIGURE 2.3.2 Simplified Flow map for the task of acceleration to a higher speed.

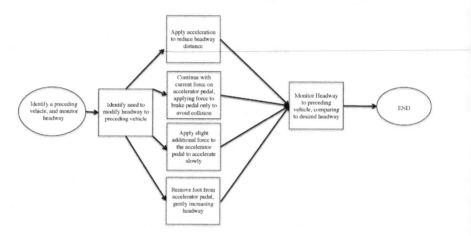

FIGURE 2.3.3 Simplified Flow map for the task of maintaining headway.

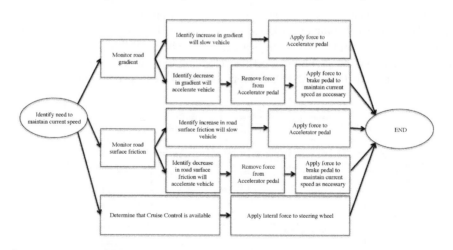

FIGURE 2.3.4 Simplified Flow map for the task of maintaining current speed.

Taking the example of Figure 2.3.1, it can be seen within this simplified flow map that once the driver has identified the need to decelerate to a lower speed, they can take three actions to achieve their goal.

1. Directly applying force to the brake pedal, allowing the vehicle to rapidly decelerate.
2. Remove force from the accelerator pedal and rely upon engine braking to gradually slow the vehicle. The driver will consequently only apply force to the brake pedal when absolutely necessary in order to maintain required headway.
3. Determine that a deceleration would not be required if the vehicle was in a different lane of the road, and acting to change the vehicle's lateral position.

For both actions whereby the vehicle maintains its current heading, the fundamental procedural requirements are consistent. The driver must remove force from the accelerator pedal and may have to, depending on distance available, apply pressure to the brake pedal. Fuel use in this context is related not to the mechanical actions that the driver takes, which due to their prescribed nature cannot significantly vary, but rather relate to the forces applied and the timings of the drivers' actions. By alerting drivers to the need to react to a potential hazard earlier, possible due to the use of V2X communication, identified within the abstraction hierarchy, drivers will be able to rely more on engine braking to slow the vehicle, reducing the need to apply force to the brake pedal, therefore mitigating the need to take further actions. This offers support for the need for context-aware interfaces and interventions that can inform the driver of upcoming actions that may be required, as identified within the CAT analysis, reducing the need for braking and improving overall fuel efficiency.

SOCIAL ORGANISATION AND COOPERATION ANALYSIS

The CWA evaluation continued to examine and complete a SOCA. SOCA seeks to identify which agents, both human and non-human, can control and influence a situation (Houghton et al., 2015). SOCA can be seen as a way to map the agents who have responsibilities within different work situations. Multiple actors can influence or constrain a situation, be this sequentially or simultaneously, supporting distributed working. Although it has been argued that SOCA can play an important role when considering initial allocation of tasks, it can also be useful for considering which tasks can be reallocated in existing systems to improve usability and efficiency, for example away from a human operator and towards automation (Naikar et al., 2006). Whilst no specific tool for SOCA has been developed (Vicente, 1999), it is widely accepted that this stage should build on previous work completed during the CWA investigation, as such for the current study SOCA was completed as an annotated CAT.

For the current CWA, nine agents were identified as playing a role in fuel-efficient driving; "Driver", "On-Board Computers", "On-Board Displays", "On-Board Sensors", "Infrastructure Network", "Other Road Users", "Vehicle Mechanical Systems", "Vehicle Electrical Systems", and "The Law". These actors were considered for each of the Object Related Processes identified within the abstraction hierarchy and for the situations identified within the CAT. The role of two additional actors was discussed amongst team members at length, that of Pedestrians and Passengers. Pedestrians were discussed in relation to the specialist nature of pedestrians in affecting road vehicles and traffic flow, however it was agreed that, for the considered situations, pedestrians could be classed as "Other Road Users" and were therefore not included in the final analysis. The role of "Passengers" was also discussed as a possible actor within the system, as passengers could influence a driver's actions, both positively, for example alerting the driver to upcoming hazards which they may be unaware of, or negatively, for example by distracting the driver and disrupting their ability to complete the driving task. It was decided, however, after considerable reflection, that the inclusion of this actor would not contribute to the analysis in any meaningful way. Passengers can influence drivers' decisions during all stages of a journey, regardless of road type or specific manoeuvre the drivers are currently engaged in, in a non-predictable way. Furthermore, passengers will not be present for all journeys a driver completes, and as such cannot be seen as a consistent constraint. The UK Department for Transport (2005), suggests that up to 89% of car journeys are single occupancy, although this figure varies considerably based on geographical location. Although passengers may influence fuel use to a limited extent, the lack of consistency and predictability meant that this category was not taken forward.

An excerpt of the competed SOCA-CAT analysis is presented in Figure 2.4. It can be seen that for some functions, only one actor is present, for example "Display Vehicle Speed" is a function solely performed by the On-Board Displays, whereas a function such as "Most Fuel-Efficient Path" is constrained by On-Board Computers, On-board Sensors, Infrastructure, Other Road Users, and The Law. The SOCA analysis highlighted the role that different agents can play in day-to-day driving fuel use and presents a myriad of opportunities for targeting future

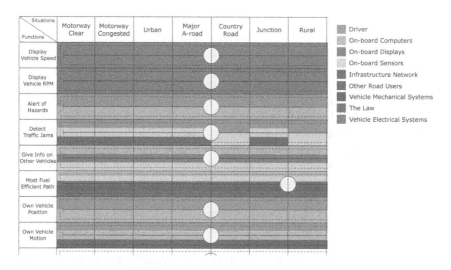

FIGURE 2.4 An excerpt from the completed SOCA-CAT analysis of the agents involved in fuel-efficient driving, including legend.

interventions. As each of the identified agents impact fuel economy to some extent, the SOCA-CAT identified the tasks whereby in-vehicle interventions would be potentially most beneficial.

WORKER COMPETENCY ANALYSIS

As the final stage of the CWA, WCA proceeds to consider the abilities and behaviour of the agents completing the work. Typically, this stage is considered using the SRK framework (Rasmussen, 1974). As such, this stage can be seen to be shifting towards the psychological constraints of users as opposed to the physical and design constraints of the system considered previously (Kilgore & St-Cyr, 2006). Vicente (1999) proposed the use of the SRK Taxonomy, which maps out the behaviour and cognitive processes that an individual may manifest when completing work-based activities. The SRK taxonomy considers that human control behaviour can be classified under three distinct levels. Skills Based Behaviour (SBB) is classified as automatic responses to environmental stimuli, often without the need for individuals' direct attention. Rules Based Behaviour (RBB) is based on linking perceptual cues to a desired outcome, following set guidelines and rules. Unlike SBB, RBB involves active decision making, allowing an individual engaged in such behaviour to vocalise their intentions and behaviour. Knowledge Based Behaviour (KBB) can be seen as the deployment of extensive and complex reasoning to find potential solutions to a problem. The use of KBB is slower and considerably more effortful than either SBB or RBB due to the need for extensive attentional focus. Tasks are not necessarily set at a given level of the taxonomy but are dependent on users' ability and experience. Novice users may be forced to rely on KBB as they are unfamiliar with a task, whereas more experienced users may too be able to use RBB as they have greater familiarity with the system and its functioning. Expert users may be able to utilise SBB for some activities should extensive training have occurred previously. Previous research has identified that the mapping of expertise to the SRK taxonomy can be highly beneficial to the development of novel training regimes (Fleming & Pritchett, 2016).

Although it is typical that the SRK taxonomy is developed using the decision ladders (McIlroy & Stanton, 2015a), due to prior work these were not considered within the current study. An alternative approach is to use the Object Related Processes layer, presented within the abstraction hierarchy (McIlroy & Stanton, 2011; Stanton & Bessell, 2014). This approach can be seen as beneficial as it directly builds on insights gained previously within the CWA, including the CAT and SOCA-CAT. Table 2.3 presents a subset of the generated SRK taxonomy using this approach. From this table, it is clear that techniques that increase drivers' skill level, operating within the SBB level, either as a result of increased experience or greater information provision, will enable users to make more environmentally conscious decisions. Further work will be required to design and develop interfaces that provide sufficient information to fulfil these requirements but do not provide excessive and/or redundant information. Previous research (Kalyuga et al., 1997) has indicated that too much information can often be just as detrimental to user performance as insufficient information.

TABLE 2.3
SRK Taxonomy

	Skill	**Rule**	**Knowledge**
Alert of Hazards	Consistently monitor the upcoming road environment and scan for potential hazards.	Monitor the current and developing road situation for potential hazards.	Follow guidance presented in highway code to maintain alertness of potential hazards.
Detecting Traffic Jams	Consistently and proactively monitor road situation for upcoming signs of traffic jams.	Monitor current road situation for traffic jams.	Follow guidance presented in highway code for identifying congested traffic.
Most Fuel-Efficient Path	Travel route which is most fuel efficient, based on previous experience and understanding of fuel economy.	If multiple paths are available, then have an awareness of which is the most fuel-efficient.	Consider the most fuel-efficient path after extensive calculation.
Shortest Path	Travel route with is the shortest geographical distance, between start and end destination.	If multiple paths are available, then have an awareness of which is the shortest based geographical distance.	Consider the shortest path between start and end destinations, based on geographical distance, after extensive calculation.
Road Gradient	Understanding relating to vehicle behaviour and response to being on a gradient and active preparation for road gradient to minimise the impact on journey and fuel efficiency.	Possess rules relating to the influence of road gradient on driving style, vehicle capabilities and fuel efficiency.	Following initial training, understand that road gradient can affect journey and fuel use.
Speed Limit	Use experience and understanding to minimise the impact changes in legal speed limit can have on journey profile and fuel economy.	Possess rules-based knowledge of speed limit.	Understand that the roadway environment contains multiple different speed limits that can affect driver actions and fuel efficiency.
Traffic Lights	Use gained understanding of how best to interact with traffic lights to minimise fuel use.	If traffic lights are present, determine speed that minimises impact of potential requirement for stopping.	Refer to legal guidelines regarding interactions with traffic lights available within highway code.
Control Acceleration	Control rate of acceleration of vehicle in a way that minimises excess fuel use.	If change of velocity is required, control rate of acceleration.	Refer to legal guidelines regarding rate of acceleration.

(Continued)

TABLE 2.3 (Continued)
SRK Taxonomy

	Skill	Rule	Knowledge
Control Vehicle Speed	Control vehicle speed using gained understanding and experience in order to minimise excess fuel use.	If vehicle is in motion, control vehicle speed.	Refer to legal guidelines regarding acceptable ranges of vehicle speed.
Control Vehicle Headway	Employ experience and understanding to control vehicle headway.	If vehicle is in motion, control vehicle headway.	Refer to initial training and legal guidelines available within highway code regarding controlling vehicle headway.
Detect Speed of Other Vehicles	Awareness of the potential importance of the speed of other vehicles on journey and fuel efficiency.	Apply rules to determine speed of other vehicles.	Understand that the speed of other vehicles can have an influence on own journey.
Smooth Motion	Full understanding regarding how smooth motion can impact driver and passenger comfort and fuel economy.	Awareness that smooth motion can impact driver passenger comfort and fuel economy.	Understand that smooth motion can impact driver passenger comfort and fuel economy.
Detect Distance of Other Vehicles	Awareness of the potential importance of the distance of other vehicles and their potential impact on fuel economy and safety.	Apply knowledge and experience to determine distance of other vehicles.	Understand that distance of other vehicles can have a direct influence on own journey.

GENERATING SPECIFICATIONS

Using the CWA approach, this chapter has identified the constraints that operate around a system designed to reduce the fuel use of day-to-day driving. Insights generated within this analysis can be used to directly inform the specifications of the required system. Championing the positive role CWA can play in devising requirement specifications is not novel and has been discussed at length by McIlroy and Stanton (2012). They propose that as the WDA stage of the CWA process, the abstraction hierarchy, offers a focus on *what* the system should achieve, independent of *how* these goals should be achieved, considerable parallels can be drawn to industrial requirement specification documents. This can be seen as an expansion on the work of Vicente (1999), who argues that the early stages of the CWA process, including the WDA, is independent of the users, their skills, and any automation that may exist within the system. Despite this potential link however, few direct examples of the use of CWA to devise system specifications exist within the literature. The work that has been conducted (Naikar & Sanderson, 1999; Naikar & Sanderson, 2001; Naikar 2006) has focussed primarily on the military domain. Taking ideas generated within CWA forward towards a considerably less prescribed domain, whereby users have considerable freedom in their actions, such as is the case when considering fuel-efficient driving is a great challenge. McIlroy and Stanton (2015a) argue that, in addition to the WDA stage, the WCA stage of CWA can provide valuable insight when designing specifications. McIlroy and Stanton (2015a) report that across the literature there is considerable variation in the individuals' use of the different stages of the CWA process in informing design. They argue however that despite a step-by-step procedure for progressing CWA not being available, indeed perhaps not even possible, the use of the CWA process as a whole can be used to elucidate design. McIlroy and Stanton (2015b) advocate that using all CWA phases is required to achieve best results, as individual phases of the analysis cannot adequately support the perspectives of designers and developers.

Taking the ideas of this chapter forward into designing suitable interfaces to support fuel-efficient driving, we can start to consider how each component of the completed CWA can inform design decisions. To start this process, it is possible to consider the different inputs a driver could make in order to control the vehicle, or directly request information from the vehicle. These considerations focussed on speech requests, direct physical control, and in-vehicle automation. Subsequently, the possible outputs a vehicle could make in response to drivers' requests, including how the vehicle can provide information and feedback to the driver, focussing on the sensory modalities of visual feedback, auditory feedback, and haptic feedback was considered. Finally, insight relating to the external environment was examined for each stage of the CWA process. This focussed on insights relating to the role of infrastructure items and a general category of "other" elements including weather conditions that could influence the driver and fuel economy. Table 2.4 presents the compiled information gained from this specification generation activity. As can be seen, not all stages of the CWA inform each of the generated input/output modalities, however, all stages offer unique insight that would not have been obtained should that stage not have been completed.

With this table in place, it is possible to identify a series of case studies whereby information or feedback could be provided to the driver. Working through Table 2.4, a series of case studies were conducted to identify the way in which the CWA could help inform specifications development. Three exemplar situations are presented in Table 2.5. As can be seen, the different stages of the CWA process provide qualitatively different insights that can be used to inform future development. Supporting previous research, the most significant information was generated within the initial WDA stage (McIlroy & Stanton, 2012; Stanton & Allison, 2020). Despite this, all stages actively contributed to the understanding of the fuel-efficient driving task (McIlroy & Stanton, 2015b). Although limited insights were gathered from the StrA stage, this can be seen as a consequence of the prescribed nature of driving, with set operations being required to achieve set goals. Should the task under examination have greater freedom in achieving goals, there is no doubt this stage of the analysis

TABLE 2.4
Specification Insights from Each CWA Stage

	Driver Input			Vehicle Output			External Environment	
	Speech	**Direct Control**	**Automation**	**Visual**	**Auditory**	**Haptic**	**Infrastructure**	**Others**
WDA (Abstraction Hierarchy)	Potential for the Driver to verbally request information -Range -Battery SoH, SoC -Journey Time -States of Traffic	Direct physical input Headway Acceleration Steering Braking	Self-Driving Control of Headway Control of Acceleration Control Route selection	Visual Displays HDD HUD	Potential for vehicle to verbally respond to queries -In-vehicle "Siri/Alexa"	Physical feedback Feedback from pedals And haptic seat	Road Network	-Traffic -V2X -V2V -Weather
CAT	When speech based information is available Where speech based information available	When set controls and information is available Where set controls and information is available	When automation would be available Where automation would be available	When visual information and feedback would be available Where visual Information and feedback would be available	When auditory information and feedback would be available Where auditory information and feedback would be available	When haptic information and feedback would be available Where haptic information and feedback would be available	Identification of road types	Identification of journey segments

(*Continued*)

TABLE 2.4 (Continued)
Specification Insights from Each CWA Stage

	Driver Input			Vehicle Output			External Environment	
	Speech	Direct Control	Automation	Visual	Auditory	Haptic	Infrastructure	Others
StrA (flow Maps)		Steps required to achieve set goals					Guidance regarding how to interact with key infrastructure elements	Guidance regarding how to interact with other roadside elements Guidance regarding how to interact with other road users
SOCA-CAT	When speech based information is available to the driver Where speech based information is available to the driver	When set controls and information is available to the driver Where set controls and information is available to the driver	When automation would be available for driver to activate Where automation would be available to activate	When visual information and feedback would be available Where visual Information and feedback would be available Who would have access to generated feedback	When auditory information and feedback would be available Where auditory information and feedback would be available Who would have access to generated feedback	When haptic information and feedback would be available Where haptic information and feedback would be available Who would have access to generated feedback		Identify possible agents, both human and non-human who can influence the fuel efficiency of the drive
WCA (SRK)		SRK guidance regarding how to achieve set goals	Rules based programmed behaviours					

TABLE 2.5
Example Case Studies regarding the Insights Offered by CWA

Scenario	WDA	CAT	StrA	SOCA	WCA
Driver wishing to learn about their vehicles remaining range, whilst maintaining a steady speed.	Vehicle must be able to present information regarding remaining range. Define how this information is to be displayed for example via the use of visual displays (HDD, HUD) or auditory channels.	Define when this information would be available to be accessed, for example road type, location, and traffic conditions.	Provide guidance on how the vehicle can maintain steady speed.	Provide information regarding who has access to vehicle range information. Potential users include the driver, on-board vehicle automated systems and potentially other road users.	Knowledge of how the vehicles range can be modified based on driver's actions, utilising previous experience.
Driver wishing to know the most fuel-efficient, route to travel from their current position to their destination.	Vehicle must have access to, and present information relating to the most fuel-efficient path, accounting for journey length, road types and traffic conditions, vehicle current position and GPS data. Define how this information is to be displayed for example via the use of visual displays (HDD, HUD) or auditory channels.	Define when this information would be available to be accessed, for example road type, location, and traffic conditions.		Provide information regarding who has access to vehicle range information. Potential users include the driver, on-board vehicle automated systems, and potentially other road users.	Knowledge of how the vehicles fuel-efficiency can be modified based on driver's actions, utilising previous experience.
Driver wishes to activate in-vehicle automation.	Vehicle must have access to some level of automation – for example basic cruise control.	Define when this information would be available to be accessed, for example road type, location, and traffic conditions.		Define the limits and potential of the automation, for example the vehicular systems the automation has access to.	Define the pre-determined behavioural rules the automation must obey for each road situation.

would offer greater insights. Although these case studies are not the focus of the current chapter, which seeks to document the development of the CWA, it is clear that the knowledge gained from a complete CWA can assist in the development of specifications for required in-vehicle interfaces across a variety of different road-based scenarios. Once fully developed, these in-vehicle interfaces can be taken forward for further empirical evaluation and testing.

CONCLUSIONS

This chapter analysed the constraints that operate around the use and development of technology supporting fuel-efficient driving by the development of a complete CWA. An initial abstraction hierarchy was presented, which documented the main objectives of the system, the metrics for judging success, considered the processes that each of these items undertook, and decomposing the physical objects that played a role in the system. Following this development, a contextual activities template was documented, decomposing the constraints, both spatial and temporal that acted on the system. Potential tasks a driver could be faced with were considered and a series of simplified flow maps constructed, examining the steps that a driver could go through in order to achieve a set target end state. In order to consider the role that different agents can take within a given scenario, a social organisation and cooperation analysis was produced considering both human and non-human agents that can influence fuel economy. Finally, a WCA was completed, using the SRK taxonomy to examine the different actions experts and novices can take when faced with everyday driving situations.

Future work will seek to build on the current analysis by developing a series of potential interfaces to support fuel-efficient driving. These interfaces will be developed with the aid of expert and novice drivers, and tested both in simulator and on-road. From the current analysis, it is argued that CWA provides a usable and accessible approach for mapping the constraints faced by users and acts as an approach that can inform future interface development and provide the basis for specification documentation.

The CWA approach brought forward a technologically agnostic tool to consider the requirements of fuel-efficient driving, highlighting the constraints that exist around this approach to driving, both in relation to the general driving task and in specific road situations. The use of CWA also allowed the identification and consideration of a variety of agents, both human and non-human, which can influence fuel use. Although the analysis highlighted that the driving task is largely prescriptive, with set mechanical operations required to control the vehicle and achieve set goals, opportunities for fuel saving interventions were identified, primarily through the use of information provision. Greater knowledge of upcoming road events can enable drivers to react earlier, encouraging a more refined driving style, and potentially reducing the amount of fuel consumed each trip. Although the relative saving of such actions is small, were these changes consistently repeated across all drivers, pollution and fuel use could drop significantly.

REFERENCES

Ahlstrom, U. (2005). Work domain analysis for air traffic controller weather displays. *Journal of Safety Research, 36*(2), 159–169.

Allison, C. K., & Stanton, N. A. (2019). Eco-driving: The role of feedback in reducing emissions from everyday driving behaviours. *Theoretical Issues in Ergonomics Science, 20*(2), 85–104. DOI:10.1080/1463922X.2018.1484967.

Barkenbus, J. N. (2010). Eco-driving: An overlooked climate change initiative. *Energy Policy, 38*(2), 762–769.

Bin, S., & Dowlatabadi, H. (2005). Consumer lifestyle approach to US energy use and the related CO_2 emissions. *Energy Policy, 33*(2), 197–208.

Birrell, S. A., Young, M. S., Jenkins, D. P., & Stanton, N. A. (2012). Cognitive work analysis for safe and efficient driving. *Theoretical Issues in Ergonomics Science, 13*(4), 430–449.

Cacciabue, P. C., & Martinetto, M. (2006). A user-centred approach for designing driving support systems: The case of collision avoidance. *Cognition, Technology & Work, 8*(3), 201–214.

Cacciabue, P. C., & Saad, F. (2008). Behavioural adaptations to driver support systems: A modelling and road safety perspective. *Cognition, Technology & Work, 10*(1), 31–39.

Carsten, O. M., & Tate, F. N. (2005). Intelligent speed adaptation: Accident savings and cost-benefit analysis. *Accident Analysis & Prevention, 37*(3), 407–416.

Cornelissen, M., Salmon, P. M., Jenkins, D. P., & Lenné, M. G. (2013). A structured approach to the strategies analysis phase of cognitive work analysis. *Theoretical Issues in Ergonomics Science, 14*(6), 546–564.

Davidsson, S., & Alm, H. (2014). Context adaptable driver information – Or, what do whom need and want when? *Applied Ergonomics, 45*(4), 994–1002.

Department for Transport, (2005). Personalised travel planning: Evaluation of 14 pilots part funded by DfT. available from https://webarchive.nationalarchives.gov.uk/20110504142036/http://www.dft.gov.uk/pgr/sustainable/travelplans/ptp/personalisedtravelplanningev577

Dietz, T., Gardner, G. T., Gilligan, J., Stern, P. C., & Vandenbergh, M. P. (2009). Household actions can provide a behavioral wedge to rapidly reduce US carbon emissions. *Proceedings of the National Academy of Sciences, 106*(44), 18452–18456.

Ehsani, M., Gao, Y., Longo, S., & Ebrahimi, K. (2018). *Modern electric, hybrid electric, and fuel cell vehicles*. CRC press. Boca Raton.

Fénix, J., Sagot, J. C., Valot, C., & Gomes, S. (2008). Operator centred design: Example of a new driver aid system in the field of rail transport. *Cognition, Technology & Work, 10*(1), 53–60.

Fidel, R., & Pejtersen, A. M. (2004). From information behaviour research to the design of information systems: The cognitive work analysis framework. *Information Research: An International Electronic Journal, 10*(1), paper 210.

Fleming, E., & Pritchett, A. (2016). SRK as a framework for the development of training for effective interaction with multi-level automation. *Cognition, Technology & Work, 18*(3), 511–528.

Gillespie, T. D. (1992). *Fundamentals of vehicle dynamics* (Vol. 114). SAE Technical Paper, Warrendale, PA, USA.

Hill, N., Brannigan, C., Smokers, R., Schroten, A., Van Essen, H., & Skinner, I. (2012). eu Transport ghg: Routes to 2050 ii. *final project report funded by the European Commission's Directorate-General Climate Action, Brussels*.

Horrein, L., Bouscayrol, A., Cheng, Y., Dumand, C., Colin, G., & Chamaillard, Y. (2016). Influence of the heating system on the fuel consumption of a hybrid electric vehicle. *Energy Conversion and Management, 129*, 250–261.

Houghton, R. J., Baber, C., Stanton, N. A., Jenkins, D. P., & Revell, K. (2015). Combining network analysis with cognitive work analysis: Insights into social organisational and cooperation analysis. *Ergonomics, 58*(3), 434–449.

Jamson, A. H., Hibberd, D. L., & Merat, N. (2015). Interface design considerations for an in-vehicle eco-driving assistance system. *Transportation Research Part C: Emerging Technologies, 58*, 642–656.

Jenkins, D. P., Stanton, N. A., Salmon, P. M., Walker, G. H., & Young, M. S. (2008). Using cognitive work analysis to explore activity allocation within military domains. *Ergonomics, 51*(6), 798–815.

Kalyuga, S., Chandler, P., & Sweller, J. (1997). Levels of expertise and user-adapted formats of instructional presentations: A cognitive load approach. In: A. Jameson, C. Paris, & C. Tasso (Eds.), *User modelling: Proceedings of the sixth international conference (UM 97)* (pp. 261–272). New York: Springer.

Kant, V. (2017). Revisiting the technologies of the old: A case study of cognitive work analysis and nanomaterials. *Cognition, Technology & Work, 19*(1), 47–71.

Kilgore, R., & St-Cyr, O. (2006, October). The SRK inventory: A tool for structuring and capturing a worker competencies analysis. In: *Proceedings of the Human Factors and Ergonomics Society Annual Meeting* (Vol. 50, No. 3, pp. 506–509). Sage CA: Los Angeles, CA: SAGE Publications.

LaClair, T. J., Verma, R., Norris, S., & Cochran, R. (2014). *Fuel economy improvement potential of a heavy-duty truck using V2x communication.* Oak Ridge, TN (United States): Oak Ridge National Lab (ORNL). National Transportation Research Center (NTRC).

Lintern, G., Cone, S., Schenaker, M., Ehlert, J., & Hughes, T. (2004). Asymmetric adversary analysis for intelligent preparation of the battlespace (A3-IPB). United States Air Force Research Department Report.

McIlroy, R. C., & Stanton, N. A. (2011). Getting past first base: Going all the way with cognitive work analysis. *Applied Ergonomics, 42*(2), 358–370.

McIlroy, R. C., & Stanton, N. A. (2012). Specifying the requirements for requirements specification: The case for work domain and worker competencies analyses. *Theoretical Issues in Ergonomics Science, 13*(4), 450–471.

McIlroy, R. C., & Stanton, N. A. (2015a). A decision ladder analysis of eco-driving: The first step towards fuel-efficient driving behaviour. *Ergonomics, 58*(6), 866–882.

McIlroy, R. C., & Stanton, N. A. (2015b). Ecological interface design two decades on: Whatever happened to the SRK taxonomy? *IEEE Transactions on Human-Machine Systems, 45*(2), 145–163.

McIlroy, R. C., Stanton, N. A., & Harvey, C. (2013). Getting drivers to do the right thing: A review of the potential for safely reducing energy consumption through design. *IET Intelligent Transport Systems, 8*(4), 388–397.

Miller, S. A., & Heard, B. R. (2016). The environmental impact of autonomous vehicles depends on adoption patterns. *Environmental Science and Technology, 50*(12), 6119–6121.

Naikar, N. (2006). Beyond interface design: Further applications of cognitive work analysis. *International Journal of Industrial Ergonomics, 36*(5), 423–438.

Naikar, N. (2013). *Work domain analysis: Concepts, guidelines, and cases.* CRC Press, Boca Raton.

Naikar, N., & Lintern, G. (2002). A review of "Cognitive work analysis: Towards safe, productive, and healthy computer-based work" by Kim J. Vicente. *The International Journal of Aviation Psychology, 12*(4), 391–400.

Naikar, N., & Sanderson, P. M. (1999). Work domain analysis for training-system definition and acquisition. *The International Journal of Aviation Psychology, 9*(3), 271–290.

Naikar, N., & Sanderson, P. M. (2001). Evaluating design proposals for complex systems with work domain analysis. *Human Factors, 43*(4), 529–542.

Naikar, N., Moylan, A., & Pearce, B. (2006). Analysing activity in complex systems with cognitive work analysis: Concepts, guidelines and case study for control task analysis. *Theoretical Issues in Ergonomics Science, 7*(4), 371–394.

Naikar, N., Pearce, B., Drumm, D., & Sanderson, P. M. (2003). Designing teams for first-of-a-kind, complex systems using the initial phases of cognitive work analysis: Case study. *Human Factors: The Journal of the Human Factors and Ergonomics Society, 45*(2), 202–217.

Oinas-Kukkonen, H., & Harjumaa, M. (2009). Persuasive systems design: Key issues, process model, and system features. *Communications of the Association for Information Systems, 24*(1), 28.

Pampel, S. M., Jamson, S. L., Hibberd, D., & Barnard, Y. (2017). The activation of eco-driving mental models: Can text messages prime drivers to use their existing knowledge and skills? *Cognition, Technology & Work, 19*(4), 743–758.

Rasmussen, J. (1986). Information Processing and Human-Machine Interaction. An Approach to Cognitive Engineering. New York: Elsevier Science Inc

Rasmussen, J. (1990). The role of error in organizing behaviour. *Ergonomics, 33*(10–11), 1185–1199.

Rasmussen, J. (1974). The Human Data Processor as a System Component: Bits and Pieces of a Model (Report No. Risø-M-1722). Danish Atomic Energy Commission, Roskilde, Denmark.

Rasmussen, J., Pejtersen, A. M., & Goodstein, L. P. (1994). *Cognitive systems engineering.* New York: Wiley.

Rasmussen, J., Pejtersen, A. M., & Schmidt, K. (1990). *Taxonomy for cognitive work analysis.* Roskilde, Denmark: Risø National Laboratory. (Risø report M-2871).

Robinette, D., Grimmer, M., Horgan, J., Kennell, J., & Vykydal, R. (2011). Torque converter clutch optimization: Improving fuel economy and reducing noise and vibration. *SAE International Journal of Engines, 4*(1), 94–105.

Sharp, T. D., & Helmicki, A. J. (1998, October). The application of the ecological interface design approach to neonatal intensive care medicine. *Proceedings of the Human Factors and Ergonomics Society Annual Meeting, 42*(3), 350–354.

Stanton, N. A., & Allison, C. K. (2020). Driving towards a greener future: An application of cognitive work analysis to promote fuel-efficient driving. *Cognition, Technology & Work, 22*(1), 125–142.

Stanton, N. A., & Bessell, K. (2014). How a submarine returns to periscope depth: Analysing complex socio-technical systems using cognitive work analysis. *Applied Ergonomics, 45*(1), 110–125.

Stanton, N. A., & McIlroy, R. C. (2012). Designing mission communication planning: The role of rich pictures and cognitive work analysis. *Theoretical Issues in Ergonomics Science, 13*(2), 146–168.

Stanton, N. A., Harris, D., & Starr, A. (2016). The future flight deck: Modelling dual, single and distributed crewing options. *Applied Ergonomics, 53*, 331–342.

Stanton, N. A., McIlroy, R. C., Harvey, C., Blainey, S., Hickford, A., Preston, J. M., & Ryan, B. (2013). Following the cognitive work analysis train of thought: Exploring the constraints of modal shift to rail transport. *Ergonomics, 56*(3), 522–540.

Strömberg, H., Karlsson, I. M., & Rexfelt, O. (2015). Eco-driving: Drivers' understanding of the concept and implications for future interventions. *Transport Policy, 39*, 48–54.

Takezaki, J., Ueki, N., Minowa, T., & Kondoh, H. (2000). Support system for safe driving. *Hitachi Review, 49*(3), 107.

Thornton, J., & Covington, H. (2015). Climate change before the court. *Nature Geoscience, 9*(1), 3–5.

Tulusan, J., Staake, T., & Fleisch, E. (2012, September). Providing eco-driving feedback to corporate car drivers: What impact does a smartphone application have on their fuel efficiency? In: *Proceedings of the 2012 ACM Conference on Ubiquitous Computing* (pp. 212–215). ACM.

Tunnell, J., Asher, Z. D., Pasricha, S., & Bradley, T. H. (2018). Toward improving vehicle fuel economy with ADAS. *SAE International Journal of Connected and Automated Vehicles*, *1*(12-01-02-0005), 81–92.

Van Westrenen, F. (2011). Cognitive work analysis and the design of user interfaces. *Cognition, Technology & Work*, *13*(1), 31–42.

Vandenbergh, M. P., & Steinemann, A. C. (2007). The carbon-neutral individual. *NYUL Rev*, *82*, 1673.

Vandenbergh, M. P., Barkenbus, J., & Gilligan, J. (2007). Individual carbon emissions: The low-hanging fruit. *UCLA L. Rev*, *55*, 1701.

Vicente, K. J. (1999). *Cognitive work analysis: Towards safe, productive, and healthy computer-based work*. Mahweh, NJ: LEA.

Wu, Y., Zhao, X., Rong, J., & Zhang, Y. (2017). How eco-driving training course influences driver behavior and comprehensibility: A driving simulator study. *Cognition, Technology & Work*, *19*(4), 731–742.

Yan, X., Fleming, J., & Lot, R. (2018). A/C energy management and vehicle cabin thermal comfort control. *IEEE Transactions on Vehicular Technology*, *67*(11), 11238–11242.

Young, M. S., Birrell, S. A., & Stanton, N. A. (2011). Safe driving in a green world: A review of driver performance benchmarks and technologies to support 'smart' driving. *Applied Ergonomics*, *42*(4), 533–539.

3 Adaptive Driver Modelling in Eco-Driving Assistance Systems

INTRODUCTION

Advanced driver assistance systems (ADAS), such as lane departure warning, curve warning, and collision warning systems, are effective in reducing the incidence and severity of road accidents. Estimates of the reduction in the number of rear-end collisions given by collision warning systems have been as high as 80% among distracted drivers, and the same system provides a safety benefit to attentive drivers by reducing the time required to release the accelerator before a potential crash (Lee et al., 2002). A lane departure warning was shown by Kozak et al. (2006) to decrease reaction time to a lane excursion by a factor of two among sleep-deprived drivers, with greater reductions when feedback was given in the form of vibration. ADAS have also been designed that promote fuel-efficient driving (Staubach et al., 2014), motivated by studies showing that feedback is necessary to retain learned eco-driving behaviours (Froehlich et al., 2009).

In practice, the efficacy of ADAS is limited by user acceptance; even the most accurate warning system is useless if the driver disables it. Operator disablement of warning systems has been observed in situations as diverse as aeroplane cockpits (Patterson, 1982) and nuclear power plant control rooms (Seminara et al., 1976). After design choices such as the distinctiveness and loudness of alarms, Sorkin (1988) implicates high false alarm rates as the most important factor limiting operator acceptance. For collision warning systems specifically, the probability of a crash given an alarm event is usually low due to the low prior probability of collisions (Parasuraman et al., 1997), so that false alarms are quite common. User acceptance of curve warning systems for cars is also known to be problematic, as shown in operational tests (LeBlanc, 2006).

To reduce the number of false alarms generated by driver assistance systems while retaining a high sensitivity to potential accidents, good understanding of driver behaviour is vital. For example, an understanding of typical car-following behaviour is needed for effective collision warnings, so that false alarms are not generated during normal conditions. But following behaviour varies greatly between drivers, with Winsum and Heino (1996) showing time headways during car-following ranging from 0.67s to 1.52s, and minimum time-to-collision values during braking varying from 2.5s to 5.1s. Likewise, to design curve warning systems that detect inattention and caution drivers to slow down, it is necessary to predict the speed at which an

DOI: 10.1201/9781003081173-3

attentive driver would take a curve. In cornering, Reymond et al. (2001) demonstrate that typical lateral accelerations differ considerably from driver to driver with an observed range of 6.4 m/s² to 11.4 m/s². In addition to this variation between drivers, the behaviour of a single driver may also change depending on road conditions and driver fatigue (Brown, 1994).

A potential solution to this large variability in driving behaviour is to design adaptive ADAS that modify their behaviour to fit the characteristics of a driver (Fleming et al., 2019). This is made possible by the increasing number of sensors on modern vehicles. One such system has already been suggested by Wang et al. (2013), which recursively updates estimates of preferred time headway and driver sensitivity to deviations in headway and inverse time-to-collision. A key problem here is to identify parameters that are consistent at different times and hence may reliably be used to characterise the driver. Updating such a model in real-time requires that we describe the driver by a small number of measurable and physically meaningful parameters.

In the present chapter, we review the existing literature on modelling of car-following and curve driving and critically evaluate these models using the results of a small-scale naturalistic driving study. The implications of this analysis for adaptive ADAS are discussed, considering what parameters of these models may be estimated by on-board sensors and the benefit in terms of user acceptance of collision and curve warning systems. Finally, naturalistic data analysis is used to recommend a set of parameters that may be used to characterise driver behaviour for adaptive ADAS. These can be used by the designers of driver assistance systems in order to improve user acceptance and therefore lead to better safety outcomes.

ADAS FOR SAFETY AND ECO-DRIVING

A total of 1,710 people lost their lives between July 2016 and June 2017 as a result of road accidents in the United Kingdom, and a further 174,790 people were injured as a result of accidents (Department of Transport, UK Government, 2017). Since 2011, the number of people killed or injured on UK roads has fluctuated with no clear trend, and more work is needed to tackle road traffic accidents and minimise their transformative and potentially terminal impact. One approach to minimising this impact is via the use of ADAS (Marchau et al., 2005).

Estimates of the potential safety implications of ADAS are promising. Collision warning systems were shown in simulator studies to reduce the number of rear-end collisions by 80% (Lee et al., 2002), while on-road studies during the EuroFOT project have confirmed the positive effects of collision warning and adaptive cruise control systems on safety and fuel economy (Benmimoun et al., 2013). Similarly, a reduction of 59% in the number of fatal accidents (Carsten et al., 2000) is claimed to be viable with the extensive use of Intelligent Speed Adaptation systems. To achieve the goal of a positive impact on safety, such systems require high acceptance by their end-users (Lindgren & Chen, 2006). This is especially true for warning systems where false alarm rates have been implicated in limiting operator acceptance (Sorkin 1988). Within ADAS specifically, the need to explore the potential acceptance of system before implementation is necessary in order to achieve a positive safety impact (Biassoni et al., 2016).

The concept of an ADAS can be extended to not only support safety but also encourage greater fuel efficiency and more environmentally friendly driving. In addition to the impact of vehicular drive trains (Chan, 2007) and mechanical systems (Vining, 2009) on fuel usage, the way a vehicle is driven can significantly influence fuel use and emissions. Research has estimated that 5–10% of fuel can be saved if drivers pursued a more fuel-efficient, economical, and environmentally friendly driving style referred to as eco-driving (Barkenbus, 2010), which is characterised by behaviours such as modest acceleration, early gear changes, minimising unnecessary braking, and driving below the speed limit. Recent analysis of naturalistic driving carried out as part of the UDRIVE project has indicated that braking, gear shifting, and velocity choice on motorways each have effects on fuel consumption of 10% or more for conventional vehicles (Heijne et al., 2017).

Despite the advantages offered by a reduction in fuel usage and emissions, previous research has found that individuals typically struggle to maintain eco-driving behaviours long term, and rather are reliant on feedback to regulate their behaviour (Lauper et al., 2015). It has also been suggested that more individuals would adopt eco-driving if they understood the impact of their current actions (Abrahamse et al., 2005). Specifically examining transportation, (Froehlich et al., 2009) and (Meschtscherjakov et al., 2009) argue that providing feedback is a cost-effective way to encourage and reinforce eco-driving practices. Based on this, ADAS has the potential to facilitate both safer and more fuel-efficient driving. Such a system has already been evaluated in a driving simulator in Staubach et al. (2014), where a 15% fuel saving was demonstrated by encouraging drivers to coast down before intersections.

MODELS OF DRIVER BEHAVIOUR

Car-Following Behaviour

Modelling drivers' behaviour when following other vehicles has been the subject of active research since at least the mid-1940s (Herrey & Herrey, 1945). Much early research was motivated by applications to traffic management and was successful in explaining emergent properties of traffic flow from the assumption that each driver behaves according to some simple rule. For example, Pipes (1953) considers the hypothesis that a driver adjusts their speed to maintain a "legal distance" to a leader vehicle given by:

$$s = s_0 + T_0 v \tag{3.1}$$

where v is the vehicle speed. When applied to a line of vehicles this implies that each driver accelerates according to the relative velocity of the preceding vehicle, which is shown to cause velocity changes in the line of traffic that propagate as a wave. These waves are often observed in real-world traffic flow, for instance at intersections when a traffic signal turns green.

Experimental evidence for a similar model of acceleration was given by Chandler et al. (1958), who proposed that the driver accelerates in proportion to the relative velocity of the vehicle ahead but with some finite reaction time, such that vehicle acceleration is given by the expression:

$$a(t+T_r) = \lambda[v_l(t) - v(t)] \qquad (3.2)$$

where v and a denote velocity and acceleration, v_l denotes the velocity of the preceding vehicle, λ is a parameter representing the sensitivity of the driver to velocity differences, and T_r is the reaction time. An interesting feature of this model is that it can show instability for sufficiently large values of λ or T_r. This instability may be the basis of further emergent properties of traffic flow such as jams or collisions (Chandler et al., 1958). A great number of subsequent works have suggested refinements and additions to these car-following models, as summarised in the review paper by Brackstone and McDonald (1999). Similar to Chandler et al. (1958), these models consist of an equation for vehicle acceleration expressing some dependence on velocity, lead vehicle velocity, and possibly the inter-vehicle spacing.

Our interest in these models is not that they may be used to simulate traffic flow (Yang & Koutsopoulos, 1996), but rather that they characterise driver behaviour by a small number of parameters that are measurable in real-time as part of an adaptive ADAS. In particular, this limits the usefulness of models that contain parameters that must be fit to data by using optimisation techniques (Kesting & Treiber, 2008) rather than having a simple physical meaning. The intelligent driver model (IDM), proposed by Treiber et al. (2000), is useful in this regard as all parameters correspond to quantities that are readily estimated from velocity and range data. These model parameters are given in Table 3.1 along with their physical interpretations. Notably, this model assumes a spacing identical to that given by (1) when moving at a steady speed behind another vehicle.

Inter-vehicle spacings and car-following behaviour may also be studied in isolation, without considering the resulting traffic behaviour. Experiments performed on a test track by Winsum and Heino (1996) suggest that each driver has a preferred time headway that they maintain in car-following situations. Similarly in Brackstone et al. (2002) it is shown that in motorway driving, human drivers allow a following distance that increases linearly with speed, corresponding to an approximately constant time separation. Further work by the same authors (Brackstone et al., 2009) considers other factors that may affect following distances and concludes that they are

TABLE 3.1
Parameters of the IDM

Parameter	Physical Interpretation
s_0	Minimum distance to leader when stationary
T_0	Minimum time headway to leader
a_{max}	Maximum desired longitudinal acceleration
b_{max}	Maximum desired longitudinal deceleration
v_0	Desired velocity

unaffected by the level of traffic flow or the road geometry, but they are affected by the type of vehicle that is being followed and may vary with time for a given driver.

From studies carried out in an early driving simulator, Todosiev (1963) noted that the acceleration and relative velocity of a driver following another vehicle shows a "limit cycle" with maximum and minimum acceptable following distances. These cycles, visible as spirals when plotting vehicle spacing against relative velocity, are also observed in real-world driving (Brackstone et al., 2002). A perceptual basis for this phenomenon has been suggested, with the limit cycle behaviour due to physiological thresholds on detection of relative velocity (Todosiev, 1963). Later studies have developed this into a complete framework for describing car-following (Leutzbach & Wiedemann, 1986). Perhaps the most widely-cited work on driver perception in car-following is Lee (1976), which advocates the use of "time-to-collision" as a predictive variable for the onset and control of braking, based on a simplified analysis of the human visual system.

Some criticism has been levelled at the existing car-following models by Boer (1999), who argued that drivers perform many tasks simultaneously and as a result are typically satisfied with a range of conditions rather than having specific preferred vehicle spacings or speeds. This idea of "satisficing" rather than "optimising" originates in Simon (1955), which suggests that humans do not attempt to make optimal decisions, and instead classify outcomes as satisfactory or unsatisfactory and act accordingly.

CORNERING BEHAVIOUR

In the literature, steering has typically been considered as a control task independent of speed control, where the driver makes steering adjustments continuously in response to deviations in road position (McRuer et al., 1977). Similar to driver response during car-following, experiments using driving simulators have shown limit cycles in the steering control of drivers when on a straight road that is consistent with the driver applying feedback subject to perceptual limits (Baxter & Harrison, 1979). This feedback is not the only aspect of steering control, as Godthelp (1985) demonstrated a precognitive aspect to steering by occluding the view of drivers during lane-change manoeuvres. These models have been refined through the application of control theory (Hess & Modjtahedzadeh, 1990) and modern developments have included consideration of the driver's neuromuscular dynamics (Pick & Cole, 2003), but they do not consider the effect of steering on the driver's speed choice, which is important to reduce false alarms in curve warning systems and to encourage coasting before curves in eco-driving assistance systems.

To model cornering speeds, Godthelp (1986) considered the role of the human visual system in cornering and showed that a quantity called "time-to-lane crossing" is kept above a threshold by drivers negotiating a curve. This model predicts that drivers will choose lower speeds on tighter curves and narrower roads. This was further elaborated in Reymond et al. (2001), which demonstrated that a minimum time-to-lane crossing is equivalent to an upper limit to the lateral acceleration of the vehicle while cornering. In experiments carried out on a test track, Reymond et al.

(2001) further showed that drivers' lateral acceleration, denoted γ, has an upper limit that decreases when travelling at higher speeds according to the expression:

$$\gamma \leq \Gamma_{max} - \Delta v^2 \qquad (3.3)$$

where Γ_{max} is a driver parameter representing the limit on lateral acceleration tolerated by the driver at low speeds, while Δ determines the decrease of this limit with speed. As the lateral acceleration experienced while driving around a corner of curvature κ is $\gamma = \kappa v^2$, the parameter Δ can be interpreted as a margin of error allowed by the driver when visually estimating the curvature of an upcoming corner. By substituting $\gamma = \kappa v^2$ into (3.3), this bound on lateral acceleration implies an upper bound on speed for a given corner given by:

$$v \leq \sqrt{\frac{\Gamma_{max}}{\kappa + \Delta}} \qquad (3.4)$$

It is interesting to note that this decrease of maximum lateral acceleration with speed cannot be explained from technical considerations of vehicle grip and handling, as the lateral acceleration at which the vehicle loses grip and skids is not dependent on speed (Gillespie, 1992).

Other models have been suggested in the literature that considers the curvature-speed relation directly. For example, Levison et al. (2007) provide an empirically-derived relationship predicting a limit on driver speed in terms of road curvature as $v \leq \alpha \kappa^{-1/4}$, where α is a driver parameter. This is equivalent to a speed-dependent lateral acceleration bound of:

$$\gamma \leq \frac{\alpha^3}{v^2} \qquad (3.5)$$

which allows for direct comparison with the model of (Reymond et al., 2001). Drawing from observed relationships between velocity and path curvature in human arm movements (Viviani & Schneider, 1991), and subsequent links to optimal control theory in the motor control literature (Viviani & Flash, 1995), Bosetti et al. (2015) suggest a "two-thirds law" relationship between the maximum speed in curves and road curvature given by $v \leq \beta \kappa^{-1/3}$ which is equivalent to a lateral acceleration bound:

$$\gamma \leq \frac{\beta^3}{v} \qquad (3.6)$$

with β the parameter characterising the driver. It has also been suggested that road width may affect driver speed choice in curves (DeFazio et al., 1992), although no account of curvature was taken in that study.

It is notable that much of the existing research into driver behaviour models, especially for following distances and cornering speeds, has been carried out on test tracks and in simulators. There is a gap in the literature in that comparatively little work has been done to validate these models in naturalistic conditions. The present study hopes to address this by starting to fill this gap.

METHODS

HARDWARE

Naturalistic driving data was collected as part of a small-scale study, carried out at the University of Southampton in the UK, in which data was collected using a non-intrusive, portable, automobile data acquisition module (ADAM) (Yan et al., 2017) that attaches to the bottom corner of a car windscreen on the passenger side using a suction cup. The device then gathers naturalistic driving data, specifically time series data of position and velocity (via GPS), acceleration and angular rotation (via integrated accelerometers) as well as inter-vehicle spacing using a pair of stereo cameras. ADAM has the advantage that it may easily be installed and removed by the participants in the study, without any modification to the vehicle. Because this process can be carried out daily by the participants themselves, it reduces the potential for demand characteristics whereby participants modify their behaviour in order to perform in a way pleasing to the researcher.

The stereo video captured from ADAM may be post-processed in order to provide the range to vehicles in front of the unit. A typical frame from the captured video, along with a headway value calculated from the stereo video cameras, can be seen in Figure 3.1.

NATURALISTIC DATA COLLECTION

Data was collected from a total of 6 participants, denoted F1–F6, who regularly drove through urban areas with heavy traffic. These participants took the device

FIGURE 3.1 Example of video data from ADAM.

for several days each and recorded their driving by installing ADAM in their own vehicles. They were not instructed to follow any particular route, instead of following their usual journeys to gather data. After discarding data recorded at night time due to difficulties in using ADAM to estimate range in darkness, 7 hours of time-series data remained to assess car-following behaviour.

To assess cornering behaviour, naturalistic data was collected from a further 3 participants, denoted C1–C3. Participant C1 had a daily commute through a rural area and contributed a total of 10 hours of data over 17 individual drives, most of which were in rural conditions where speed was not limited by traffic. Participants C2 and C3 contributed approximately an hour of data each using ADAM, again in rural conditions. Time-series data of GPS position from these participants was processed to provide an estimate of road curvature, and the resulting speed and acceleration data was filtered to remove noise. All participants (F1–6, C1–3) were frequent drivers aged between 25 and 40 with at least 3 years of experience, driving vehicles with engine sizes between 1.2 and 2.0 litres.

RESULTS

CAR-FOLLOWING

Initial analysis of the data revealed the limit cycles reported by Todosiev (1963) when the data is plotted in the "phase space" of vehicle spacing versus relative velocity, which is defined such that positive values correspond to the participant's car travelling faster than the leading vehicle. A typical such spiral is shown in Figure 3.2 for a slow-moving car-following situation (mean velocity

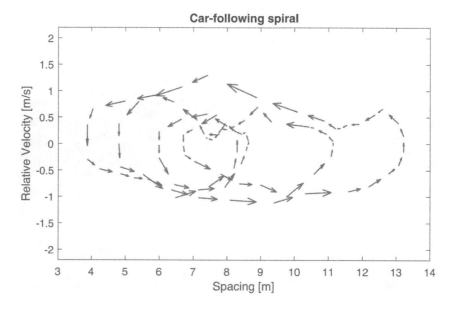

FIGURE 3.2 A typical car-following spiral.

Adaptive Driver Modelling in Eco-Driving Assistance Systems

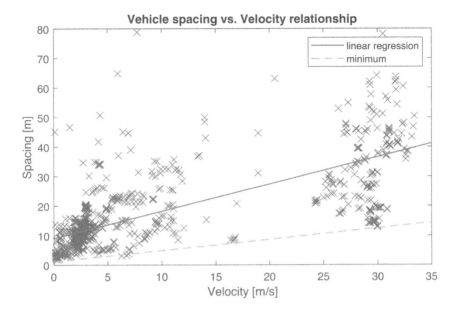

FIGURE 3.3 Observed vehicle spacings for F1.

1.5 m/s), with 60 seconds of data shown. The longitudinal accelerations of the drivers studied ranged between 4.5 m/s² and 4 m/s², with no emergency braking events observed.

We investigated the relationship between velocity and vehicle spacing by filtering and down-sampling the collected time-series data to a sample rate of 1s and retaining only those time instants where the absolute value of acceleration of both the participant's vehicle and the preceding vehicle was less than 0.5 m/s² (that is, both vehicles were travelling at approximately constant speed). The retained values of observed spacing are shown in Figure 3.3 for participant F1. There was a large variability in observed vehicle spacings for all participants, although a general trend of increasing spacing with increasing speed was observed. Figure 3.3 also shows the line of best fit obtained by linear regression, corresponding to the relationship (1), which gave $s_0 = 9.4$ m and $T_0 = 0.90$ s in this case. The wide variability in observed spacings and hence poor predictive value of this relationship is evident from the figure, with $R^2 = 0.51$.

For all participants, we also observed that the lower bound of the observed spacings increased with velocity and was well-approximated by a linear relationship as in (1). An estimate of this lower bound on observed spacing is shown in Figure 3.3 for participant F1, and the estimated values of s_{min} and T_{min} for the other participants are given in Table 3.2. The lack of points between 15 m/s and 25 m/s in the figure is due to steady following at these velocities rarely being observed in the naturalistic data. This is likely because the participants mostly drove on roads with speed limits of 30, 40, and 70 miles/hour.

To investigate the time-to-collision (TTC) in car-following, we considered the relationship between spacing and relative velocity. The naturalistic time-series was

TABLE 3.2
Observed Minimum Car-Following Parameters

Participant	TTC_{min} [s]	s_{min} [m]	T_{min} [s]
F1	2.40	1.34	0.38
F2	3.38	1.92	0.48
F3	3.13	1.45	0.45
F4	1.90	0.49	0.30
F5	3.21	1.44	0.64
F6	2.52	1.80	0.34
Mean	2.76	1.41	0.43
S. d.	±0.57	±0.50	±0.12

again filtered and down-sampled to a rate of 1s, and all-time instants when the participant car had a velocity of zero were removed. The result of this procedure is shown in Figure 3.4 for participant F1. From the figure, it is immediately apparent that there is a lower and upper bound to the observed relative velocities that appear to be well-approximated by a linear relationship. Noting that the inverse of TTC corresponds to the slope of a line on this graph, the upper bound is consistent with the driver maintaining a TTC of greater than 2.4s.

Minimum observed TTC values for the other study participants are shown in Table 3.2. We note also that the lower bound to observed TTC as shown in Figure 3.4

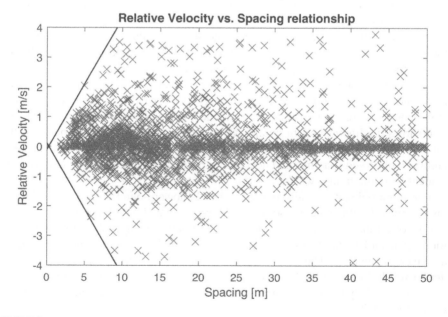

FIGURE 3.4 Observed relative velocities for F1.

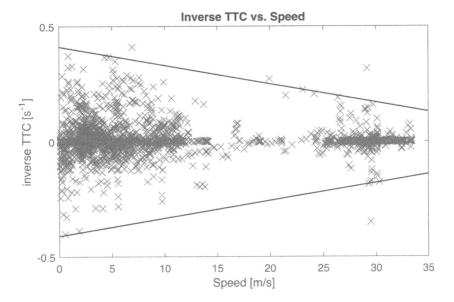

FIGURE 3.5   Observed inverse TTC for F1.

for participant F1, and the approximately linear lower bound on observed spacing shown in Figure 3.3, is observed in the data collected from the other study participants.

Finally, we investigated the possible dependence of TTC on speed by computing the inverse TTC, as shown in Figure 3.5. This figure shows the inverse TTC and speed observed for F1 at different time instants. Generally, the maximum value of inverse TTC appears to decrease with speed (corresponding to a minimum TTC that increases with speed). There also appears to be a lower bound on the inverse TTC, increasing with speed. From inspection of the figure, it appears that there is some symmetry between the upper and lower bounds, and this is was also observed for the other participants.

Cornering

The time-series data from participants C1–C3 were analysed in order to obtain relationships involving driver speed choice when cornering. All of the local maxima of curvature in the time series data were found, and for each curvature value, the corresponding observed velocity was extracted. This procedure gives a total of 7384 cornering events for C1, 657 for C2, and 730 for C3. These "corners" included junctions and instances in which the speed of the vehicle was limited by traffic flow, the effect of which is to reduce the vehicle velocity to less than that attributable to cornering alone. The lateral acceleration was then estimated from the formula $\gamma = \kappa v^2$, which is the lateral acceleration when following a curved path, and the peak value of lateral acceleration in each corner extracted from the time-series. This provides a large set of observed lateral accelerations against velocity as shown in Figure 3.6.

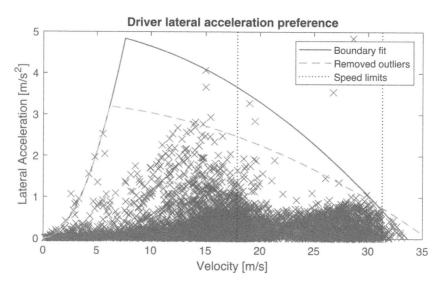

FIGURE 3.6 Observed lateral accelerations in cornering data.

Following the procedure in Reymond et al. (2001), a "high-velocity" upper bound curve following the relationship (3.3), and a "low-velocity" curve from $\gamma = \kappa v^2$, were fitted to this data with some of the observed points classified as outliers and ignored. The low-velocity part of the bound is obtained assuming a minimum turning radius for the vehicle of 10m. Depending on the number of points designated as outliers, two possible curves are obtained with values of $\Gamma_{max} = 5.1$ and $\Delta = 4.5$ or $\Gamma_{max} = 3.3$ and $\Delta = 2.6$. It is evident that more corners have been observed with velocities of between 15 m/s and 20 m/s and also 25–30 m/s than at other velocities. This uneven distribution of velocities is likely due to speed limits on the rural route that was most often driven by the participant. In particular, speed limits of 40 and 70 miles per hour were the most common, and we have indicated these values with vertical lines in Figure 3.6. The high-speed part of this plot is shown in greater detail in Figure 3.7. Qualitatively, it appears from the latter figure that the red curve with values of $\Gamma_{max} = 5.1$ and $\Delta = 4.5$ fits the data better at high speed, but this may be due to the relatively small number of corners observed above the speed limit of 70 mph (corresponding to 31.3 m/s).

These lateral acceleration limits imply an upper bound to velocity that depends on curvature as shown in Figure 3.8, which shows the velocities observed at the point of maximum curvature in each corner. It is notable that the outliers appear much closer to the rest of the data when plotting the velocity limit instead of the lateral acceleration. Also shown is a vertical line denoting the maximum recommended curvature for a 60 mph road according to the UK design guidelines for road geometry (Highways Agency, UK Government, 2002). We conjecture that this causes the rapid apparent drop off in observed velocities visible in the figure near a curvature of 1.4 rad/km, as curvatures of greater than this are only likely on roads with a speed limit of 50 mph or less due to the design codes (Highways

Adaptive Driver Modelling in Eco-Driving Assistance Systems 61

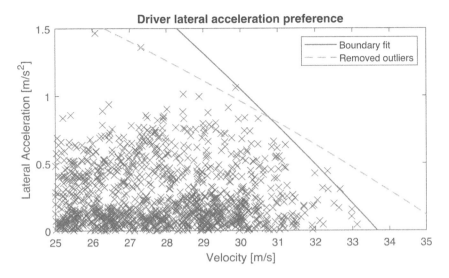

FIGURE 3.7 Detail of high-speed data.

Agency, UK Government, 2002). This may also explain the low values of peak lateral acceleration observed in Figure 3.6 between the speeds of 20 m/s and 25 m/s.

Other models of driver speed choice in curves were also evaluated against the gathered data. In particular, we assessed the "two-thirds law" of Bosetti et al. (2015) and the power law relationship of Levison et al. (2007), both of which were suggested as models to predict driver speed in curves. Fits of these two relationships as upper bounds to the observed data are shown in Figure 3.9. Levison's relationship

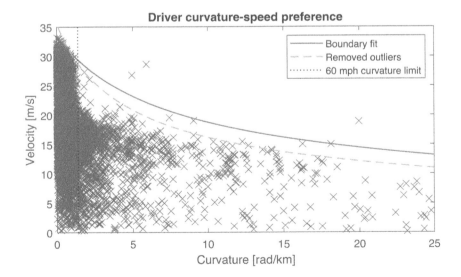

FIGURE 3.8 Observed velocities for cornering data.

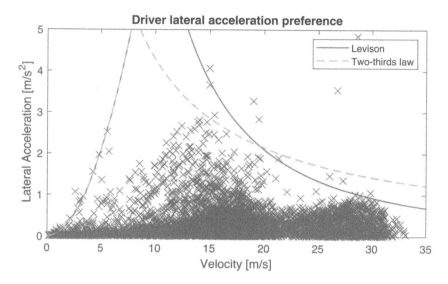

FIGURE 3.9 Alternative lateral acceleration models.

(Levison et al., 2007) appears to overestimate lateral acceleration for the corners with speeds of around 15 m/s observed in the study. Conversely, the two-thirds law (Bosetti et al., 2015) appears to overestimate the allowable lateral acceleration at high speeds, such as for the numerous 30 m/s corners seen in the data. Both of these models appear to allow very large lateral accelerations at speeds of around 10 m/s, such that they appear unlikely to provide a reasonable description of driver cornering speeds for large curvatures.

The boundary fitting procedure was repeated for participants C2 and C3 to find the corresponding lateral acceleration bounds according to (3.3), and the resulting parameters are shown in Table 3.3. It is notable that the values of the parameter Γ_{max} appear to be lower across our three participants C1–C3 than those given by Reymond et al. (2001), where the quoted mean was 7.64. A one-sample t-test for the null hypothesis $\mu = 7.64$ yields $t(2) = -4.93$, $p < 0.05$, suggesting significance, and the corresponding 95% confidence interval for the mean is [3.07, 7.27].

TABLE 3.3
Observed Cornering Parameters

Participant	Γ_{max} [m/s^2]	Δ [rad/km]
C1	5.13	4.53
C2	5.34	4.95
C3	6.03	3.99
Mean	5.98	4.49
S. d.	±0.58	±0.49

DISCUSSION

COMPARISON OF MODELS WITH NATURALISTIC DATA

Preferred vehicle spacings were observed to vary widely, even when considering the same driver at different times. This conflicts with the results of some previous studies such as Winsum & Heino (1996) which showed that drivers were consistent in their choice of headway, but this disparity may appear because that study was not done under naturalistic conditions. In particular, participants in Winsum & Heino (1996) were instructed to drive "as if they had to reach their destination as soon as possible, without overtaking other vehicles", which likely encouraged them to drive closer to the preceding vehicle. On the other hand, Brackstone et al. (2009) showed that for passive subjects who were unaware they were being studied, drivers were inconsistent in their choice of following distance. Brackstone et al. (2009) also observed a day-to-day variation in headway between the active participants in the study, supporting our conclusion that preferred following distances may vary widely for a particular driver.

A consequence is that many of the car-following models originating in the traffic literature, although successful at describing the bulk aspects of traffic flow, may be too prescriptive to accurately represent the behaviour of an individual driver. Many of these models, such as the Intelligent Driver Model (Treiber et al., 2000), include a preferred spacing for a given speed that will be approached in steady-state car following. Yet in our data, we observed that individual drivers may remain at a steady speed at a wide variety of distances (Figure 3.3). We also observed the limit cycles described by Todosiev (1963) in the car-following data (Figure 3.2), further suggesting a lack of a constant steady-state headway value. This has immediate implications for the design of adaptive collision warning systems, as it implies that it will be difficult to obtain a consistent estimate of time headway as attempted by Wang et al. (2013). We conjecture that driver behaviour is more accurately modelled if a minimum acceptable spacing is specified, which depends upon speed as in (3.1).

In contrast, when considering relative velocities there appears to be clearly defined minimum acceptable time-to-collision values (Figure 3.4), indicating that the onset of braking should be predictable according to the perceptual model of Lee (1976). Calculation of the inverse TTC reveals a dependence of the inverse TTC on speed, with an upper bound that decreases with speed, in agreement with the model in Kiefer et al. (2005). We also observed an increasing lower bound on inverse TTC with speed that corresponds to the driver accelerating to catch up to the vehicle in front of them.

Of the cornering models considered, the lateral acceleration margin model of Reymond et al. (2001) appears to fit the collected naturalistic data better than Levison's model (Levison et al., 2007) and the two-thirds law (Bosetti et al., 2015). Reymond's model also has the advantage of a well-explained perceptual basis in that drivers allow a margin for error in their estimate of curvature when cornering. However, the are several quite subjective aspects to the model fitting, such as the elimination of outliers and hence the choice of points on the boundary when performing a regression analysis. For adaptive curve warning systems, we may infer that driver preferences are measurable if it is possible to reliably identify these outliers. A further complication is that design guidelines for road geometry affect the shape of the boundary and any adaptive ADAS should not fit the resulting artificially low values of cornering speed that result from this (for instance, the region around 2–3 rad/km in Figure 3.8). These driver lateral acceleration preferences are also relevant for occupant comfort in autonomous driving, where user acceptance is likely to be low if the driver perceives the car to be going too fast or too slow in corners.

PARAMETERS CHARACTERISING DRIVER BEHAVIOUR

Generally, the most successful models in our analysis are those that prescribe bounds on quantities, such as the lateral acceleration bound of Reymond et al. (2001) or the time-to-collision bound of Lee (1976), rather than those giving expected values, such as the preferred headway of Winsum and Heino (1996). This fits with the idea of a driver "satisficing", that is ensuring satisfaction of constraints rather than optimising a performance criterion (Boer, 1999). Knowledge of these constraints is essential to understand typical driver behaviour, and hence for user acceptance of ADAS. A suggested set of parameters to be used to characterise a driver for adaptive ADAS is given in Table 3.4, and we now briefly justify these choices considering curve warning and collision warning systems.

The inverse time-to-collision (TTC) may be a more useful measure than time-to-collision, as in the data there appears to be an upper and lower bound to this quantity that may reduce in magnitude as vehicle velocity increases. Inverse TTC has already been suggested for use in forward collision warning systems, for example by Kiefer et al. (2005), and was also suggested as an input to an adaptive collision warning system (Wang et al., 2013). The observation that TTC appears to have a dependence on speed is consistent with the results in Kiefer et al. (2005).

TABLE 3.4
Recommended Parameters for an ADAS Driver Model

Parameter	Typical Value	Physical Interpretation
$iTTC_{min}$	$-0.2 s^{-1}$	Minimum inverse time-to-collision to leading vehicle
$iTTC_{max}$	$0.2 s^{-1}$	Maximum inverse time-to-collision to leading vehicle
s_{min}	2 m	Minimum distance to leading vehicle when stationary
T_{min}	0.7 s	Minimum time headway to leading vehicle
a_{max}	4 m/s²	Maximum desired longitudinal acceleration
b_{max}	5 m/s²	Maximum desired longitudinal deceleration
Γ_{max}	6 m/s²	Maximum tolerable lateral acceleration
Δ	4 rad/km⁻¹	Driver curvature safety margin

It is possible that the minimum observed spacing is consistent for each driver, as illustrated in Figure 3.3. However, the concept of a preferred spacing is in general not borne out by the naturalistic data, with linear regressions typically giving poor correlations of $R^2 \approx 0.5$. This has interesting implications for forward collision warning systems specifically, as the desired inter-vehicle spacing is not predictable from the current velocity and if using low values of headway to trigger warnings, care should be taken in the design not to sound false alarms. Notably, the adaptive ADAS of (Wang et al., 2013) uses this concept of a preferred spacing, despite the large variability we observed in this quantity during naturalistic driving. A much more consistent measure in our data is the minimum TTC (or equivalently, inverse TTC), which appears to have a well-defined acceptable range, and this quantity should be preferred when designing collision warning systems.

Naturalistic data on cornering speeds matches the model given in Reymond et al. (2001) once allowance has been made for the limits of the road geometry, and this gives either an upper bound to the lateral acceleration or equivalently an upper bound to speed as a function of curvature that should be considered in the design of curve-warning systems. For our participants the observed values of the maximum lateral acceleration parameter Γ_{max} are significantly ($p < 0.05$) lower than those given in Reymond et al. (2001), giving some evidence that the values in that paper may be too high for the naturalistic driving studied. It is possible that this represents a difference between naturalistic driving and driving on a test track as in Reymond et al. (2001), although our sample size is also very small and it could represent differences in the participants used in the study. Given that the parameters defining this bound show considerable variation between drivers, a curve-warning system should adapt to observed values of lateral acceleration to minimise false alarms and improve operator acceptance. This seems especially important as operational tests of current curve warning systems have shown limited user acceptance (LeBlanc, 2006).

Limitations

The main limitation of the present work is the small number of participants used to gather the naturalistic data. As a result of this, the numerical values of the driving parameters considered in Tables 3.4 and 3.3 should not be considered representative of drivers as a whole. Rather, the main value lies in the investigation of the intra-driver variation of quantities such as following distance and cornering velocity, and identification of the models in the existing literature that give a good description of the behaviour of individual participants. We also note that all vehicle following data was collected in the daytime and that different models may apply during hours of darkness.

The GPS unit of ADAM has a manufacturer-stated positional error of ±0.9 m and a velocity error of ±0.1 m/s. A "worst-case" error analysis of the error when estimating road curvature based on these values gives ±0.177 rad/km for a corner of curvature 10 rad/km, although in practice the error is likely to be smaller as the curvature is estimated from several consecutive position, speed, and heading measurements. Similarly, the lateral acceleration estimation error is ±0.070 m/s² when travelling at 15 m/s around a corner of curvature 10 rad/km. The distance estimation algorithm of

ADAM used to calculate inter-vehicle spacing has a range error of less than 0.1 m and a relative velocity error of less than 0.1 m/s for distances of up to 40 m, which were determined by the authors by comparison with data from a Doppler radar installed on an instrumented vehicle.

CONCLUSIONS

Motivated by the high efficacy (Lee et al., 2002; Kozak et al., 2006) but low reported user acceptance (Parasuraman et al., 1997; LeBlanc, 2006) of some driver assistance systems, we have advocated the design of ADAS that adapt to the characteristics of a driver. In doing so, the number of false alarms can be decreased for warning systems, which are a major cause of user disablement (Sorkin, 1988). But to achieve this, accurate understanding of driver behaviour in naturalistic conditions is crucial.

Using naturalistic data collected as part of a study at the University of Southampton, we evaluated the usefulness of existing models in describing observed relationships in the data during car-following and cornering, with a focus on following distance, time-to-collision, cornering speed, and lateral acceleration. In particular:

- There is a poor observed correlation between speed and inter-vehicle spacing. Speed appears to be a poor predictor for a driver's following distance, with drivers showing a large variation in their following distances at a given speed in the naturalistic data. This calls into question the use of models of this type in the design of collision-warning systems.
- Inverse time-to-collision (TTC) has a well-defined lower and upper bound for each driver that reduces in magnitude at higher speeds. This may be used to predict the onset of braking and acceleration in following situations, a finding that reinforces its use to identify critical situations for collision-warning.
- Speed while cornering is well-described by the lateral acceleration margin model of Reymond et al. (2001), which gives a better description of tolerable lateral accelerations than competing models and is a good candidate for use in curve warning and eco-driving assistance systems, although values of the model parameters appear lower in our data than in the original study.

Finally, we note that our analysis of the naturalistic data supports the notion that a driver satisfices rather than optimises when driving (Boer, 1999). For adaptive ADAS, we have therefore advocated a modelling approach using parameters that describe limits to quantities such as time headway, time-to-collision, and lateral acceleration. This will allow future driver assistance systems to adapt to driver preferences, improving user acceptance and hence improving safety outcomes.

REFERENCES

Abrahamse, W., Steg, L., Vlek, C., & Rothengatter, T. (2005). A review of intervention studies aimed at household energy conservation. *Journal of Environmental Psychology, 25*(3), 273–291.

Barkenbus, J. N. (2010). Eco-driving: An overlooked climate change initiative. *Energy Policy, 38*(2), 762–769.
Baxter, J., & Harrison, J. Y. (1979). A nonlinear model describing driver behavior on straight roads. *Human Factors, 21*(1), 87–97.
Benmimoun, M., Pütz, A., Zlocki, A., & Eckstein, L. (2013). Eurofot: Field operational test and impact assessment of advanced driver assistance systems: Final results. In: *Proceedings of the FISITA 2012 World Automotive Congress*, pp. 537–547. Springer.
Biassoni, F., Ruscio, D., & Ciceri, R. (2016). Limitations and automation: The role of information about device-specific features in ADAS acceptability. *Safety Science, 85*, 179–186.
Boer, E. R. (1999). Car following from the drivers perspective. *Transportation Research Part F: Traffic Psychology and Behaviour, 2*(4), 201–206.
Bosetti, P., Da Lio, M., & Saroldi, A. (2015). On curve negotiation: From driver support to automation. *IEEE Transactions on Intelligent Transportation Systems, 16*(4), 2082–2093.
Brackstone, M., & McDonald, M. (1999). Car-following: A historical review. *Transportation Research Part F: Traffic Psychology and Behaviour, 2*(4), 181–196.
Brackstone, M., Sultan, B., & McDonald, M. (2002). Motorway driver behaviour: Studies on car following. *Transportation Research Part F: Traffic Psychology and Behaviour, 5*(1), 31–46.
Brackstone, M., Waterson, B., & McDonald, M. (2009). Determinants of following headway in congested traffic. *Transportation Research Part F: Traffic Psychology and Behaviour, 12*(2), 131–142.
Brown, I. D. (1994). Driver fatigue. *Human Factors, 36*(2), 298–314.
Carsten, O., Fowkes, M., & Tate, F. (2000). Implementing intelligent speed adaptation in the UK: Recommendations of the EVSC project. In: *Proceedings of the 7th World Congress on Intelligent Transport Systems, ITS Congress Association, Brussels*.
Chan, C. C. (2007). The state of the art of electric, hybrid, and fuel cell vehicles. *Proceedings of the IEEE, 95*(4), 704–718.
Chandler, R. E., Herman, R., & Montroll, E. W. (1958). Traffic dynamics: Studies in car following. *Operations Research, 6*(2), 165–184.
DeFazio, K., Wittman, D., & Drury, C. (1992). Effective vehicle width in self-paced tracking. *Applied Ergonomics, 23*(6), 382–386.
Department of Transport, UK Government. (2017). Reported road casualties in Great Britain: Quarterly provisional estimates year ending June https://www.gov.uk/government/uploads/system/uploads/attachment_data/file/654962/quarterly-estimates-april-to-june-2017.pdf, 2017. Accessed: 11/12/2017.
Fleming, J. M., Allison, C. K., Yan, X., Lot, R., & Stanton, N. A. (2019). Adaptive driver modelling in adas to improve user acceptance: A study using naturalistic data. *Safety Science, 119*, 76–83.
Froehlich, J., Dillahunt, T., Klasnja, P., Mankoff, J., Consolvo, S., Harrison, B., & Landay, J. A. (2009). Ubigreen: Investigating a mobile tool for tracking and supporting green transportation habits. In: *Proceedings of the SIGCHI Conference on Human Factors in Computing Systems*, pp. 1043–1052. ACM.
Gillespie, T. D. (1992). *Fundamentals of vehicle dynamics*. SAE International.
Godthelp, H. (1986). Vehicle control during curve driving. *Human Factors, 28*(2), 211–221.
Godthelp, J. (1985). Precognitive control: Open-and closed-loop steering in a lane-change manoeuvre. *Ergonomics, 28*(10), 1419–1438.
Heijne, V., Ligterink, N., & Stelwagen, U. (2017). Potential of eco-driving. UDRIVE Deliverable D45.1. EU FP7 Project UDRIVE Consortium. https://doi.org/10.26323/UDRIVE_D45.1.
Herrey, E. M., & Herrey, H. (1945). Principles of physics applied to traffic movements and road conditions. *American Journal of Physics, 13*(1), 1–14.

Hess, R., & Modjtahedzadeh, A. (1990). A control theoretic model of driver steering behavior. *IEEE Control Systems Magazine*, *10*(5), 3–8.

Highways Agency, UK Government. (2002). Design manual for roads and bridges: Volume 6 road geometry. http://www.standardsforhighways.co.uk/ha/standards/dmrb/vol6/section2/td4295.pdf, Accessed: 11/12/2017.

Kesting, A., & Treiber, M. (2008). Calibrating car-following models by using trajectory data: Methodological study. *Transportation Research Record*, *2088*(1), 148–156.

Kiefer, R. J., LeBlanc, D. J., & Flannagan, C. A. (2005). Developing an inverse time-to-collision crash alert timing approach based on drivers last-second braking and steering judgments. *Accident Analysis & Prevention*, *37*(2), 295–303.

Kozak, K., Pohl, J., Birk, W., Greenberg, J., Artz, B., Blommer, M., Cathey, L., & Curry, R. (2006). Evaluation of lane departure warnings for drowsy drivers. In: *Proceedings of the human factors and ergonomics society annual meeting*, volume 50, pp. 2400–2404. Los Angeles, CA: Sage Publications Sage CA.

Lauper, E., Moser, S., Fischer, M., Matthies, E., & Kaufmann-Hayoz, R. (2015). Psychological predictors of eco-driving: A longitudinal study. *Transportation Research Part F: Traffic Psychology and Behaviour*, *33*, 27–37.

LeBlanc, D. (2006). *Road departure crash warning system field operational test: methodology and results. volume 1: technical report*. University of Michigan, Ann Arbor, Transportation Research Institute.

Lee, D. N. (1976). A theory of visual control of braking based on information about time-to-collision. *Perception*, *5*(4), 437–459.

Lee, J. D., McGehee, D. V., Brown, T. L., & Reyes, M. L. (2002). Collision warning timing, driver distraction, and driver response to imminent rear-end collisions in a high-fidelity driving simulator. *Human Factors*, *44*(2), 314–334.

Leutzbach, W., & Wiedemann, R. (1986). Development and applications of traffic simulation models at the karlsruhe institut für verkehrswesen. *Traffic Engineering & Control*, *27*(5), 270–278.

Levison, W. H., Campbell, J. L., Kludt, K., Bittner, A. C., Potts, I. B., Harwood, D. W., ... & Schreiner, C. S. (2007). *Development of a Driver Vehicle Module (DVM) for the Interactive Highway Safety Design Model (IHSDM) (No. FHWA-HRT-08-019)*. United States. Federal Highway Administration. Office of Research and Technology Services.

Lindgren, A., & Chen, F. (2006). State of the art analysis: An overview of advanced driver assistance systems (ADAS) and possible human factors issues. *Human factors and economics aspects on safety*, *38*, 50.

Marchau, V., Van der Heijden, R., & Molin, E. (2005). Desirability of advanced driver assistance from road safety perspective: The case of ISA. *Safety Science*, *43*(1), 11–27.

McRuer, D. T., Allen, R. W., Weir, D. H., & Klein, R. H. (1977). New results in driver steering control models. *Human Factors*, *19*(4), 381–397.

Meschtscherjakov, A., Wilfinger, D., Scherndl, T., & Tscheligi, M. (2009). Acceptance of future persuasive in-car interfaces towards a more economic driving behaviour. In: *Proceedings of the 1st International Conference on Automotive User Interfaces and Interactive Vehicular Applications*, pages 81–88. ACM.

Parasuraman, R., Hancock, P. A., & Olofinboba, O. (1997). Alarm effectiveness in driver-centred collision-warning systems. *Ergonomics*, *40*(3), 390–399.

Patterson, R. D. (1982). *Guidelines for auditory warning systems on civil aircraft*. Civil Aviation Authority, UK. Paper 82017.

Pick, A., & Cole, D. (2003). Neuromuscular dynamics and the vehicle steering task. *The Dynamics of Vehicles on Roads and on Tracks*, *41*, 182–191.

Pipes, L. A. (1953). An operational analysis of traffic dynamics. *Journal of Applied Physics*, *24*(3), 274–281.

Reymond, G., Kemeny, A., Droulez, J., & Berthoz, A. (2001). Role of lateral acceleration in curve driving: Driver model and experiments on a real vehicle and a driving simulator. *Human Factors*, *43*(3), 483–495.

Seminara, J. L., Gonzalez, W. R., & Parsons, S. O. (1976). *Human factors review of nuclear power plant control room design. Summary report* (No. EPRI-NP-309-SY). Lockheed Missiles and Space Co., Sunnyvale, CA (USA).

Simon, H. A. (1955). A behavioral model of rational choice. *The Quarterly Journal of Economics*, *69*(1), 99–118.

Sorkin, R. D. (1988). Why are people turning off our alarms? *The Journal of the Acoustical Society of America*, *84*(3), 1107–1108.

Staubach, M., Schebitz, N., Köster, F., & Kuck, D. (2014). Evaluation of an eco-driving support system. *Transportation Research Part F: Traffic Psychology and Behaviour*, *27*, 11–21.

Todosiev, E. P. (1963). *The action point model of the driver-vehicle system*. PhD thesis, The Ohio State University.

Treiber, M., Hennecke, A., & Helbing, D. (2000). Congested traffic states in empirical observations and microscopic simulations. *Physical Review E*, *62*(2), 1805.

Vining, C. B. (2009). An inconvenient truth about thermoelectrics. *Nature Materials*, *8*(2), 83–85.

Viviani, P., & Flash, T. (1995). Minimum-jerk, two-thirds power law, and isochrony: Converging approaches to movement planning. *Journal of Experimental Psychology: Human Perception and Performance*, *21*(1), 32.

Viviani, P., & Schneider, R. (1991). A developmental study of the relationship between geometry and kinematics in drawing movements. *Journal of Experimental Psychology: Human Perception and Performance*, *17*(1), 198–218.

Wang, J., Zhang, L., Zhang, D., & Li, K. (2013). An adaptive longitudinal driving assistance system based on driver characteristics. *IEEE Transactions on Intelligent Transportation Systems*, *14*(1), 1–12.

Winsum, W., & Heino, A. (1996). Choice of time-headway in car-following and the role of time-to-collision information in braking. *Ergonomics*, *39*(4), 579–592.

Yan, X., Fleming, J., Allison, C., & Lot, R. (2017). Portable automobile data acquisition module (ADAM) for naturalistic driving study. In: *Proceedings of the 15th European Automotive Congress*.

Yang, Q., & Koutsopoulos, H. N. (1996). A microscopic traffic simulator for evaluation of dynamic traffic management systems. *Transportation Research Part C: Emerging Technologies*, *4*(3), 113–129.

4 Taming Design with Intent Using Cognitive Work Analysis

INTRODUCTION

Transport emissions, primarily carbon dioxide (CO_2) and nitrous oxides (NOx), are the leading cause of air pollution in Britain (Department for Environment Food and Rural Affairs, 2017). As seen over the preceding chapters, road vehicles, specifically automobiles, are the biggest contributor to this statistic, accounting for approximately 75% of the total transport-related greenhouse gas emissions (Hill et al., 2012). Vehicle emissions are far from benign, having significant long-term and universally negative health implications (McCubbin & Delucchi, 1999). Exposure to vehicle emissions increases individuals' risk of developing a variety of respiratory disorders including asthma, bronchitis, chronic obstructive pulmonary disease, pneumonia, and upper respiratory tract infection (Buckeridge et al., 2002). In addition to the considerable negative impact vehicle emissions can have on human health, these emissions also have significant environmental impact, and have been directly linked to anthropogenic climate change and changing global weather patterns (Karl & Trenberth, 2003; Chapman, 2007). With such significant and universally negative effects, finding ways to reduce the current high levels of vehicle emissions is a defining challenge of the 21st century, one which the automotive sector is keen to address (Li et al., 2018; Allison & Stanton, 2019; Chapter 1).

Whilst it is, at least currently, not possible to completely remove emissions from automotive transportation (Chan, 2007), As championed within previous chapters, emissions and the associated volume of fuel used can be significantly reduced as a consequence of driver behaviour change (Barth & Boriboonsomsin, 2009; Barkenbus, 2010). Previous research has suggested that vehicle emissions could be reduced by 5% to 20% (Stillwater & Kurani, 2013) with fuel usage reduced between 5% and 10% (Martin et al., 2012) should drivers engage in more environmentally friendly driving behaviours. As it has been previously argued, *"There is little innately special about more environmentally friendly user behaviour: it's often simply about using a system effectively"* (Lockton et al., 2009). Pursuing interventions to support a shift towards such driving behaviour and encouraging the adoption of more environmentally conscious driving styles is therefore justly warranted.

As introduced previously, one approach to support the modification of driver behaviour is the design of interfaces that directly offer guidance on potential future actions and offers feedback on previous behaviours (McIlroy et al., 2013). The design of new interfaces to encourage a greater awareness of resource use is not novel and has been significantly pursued to reduce both household energy usage (Jain et al., 2012) and vehicle energy usage whilst driving (Jamson et al., 2015). Feedback devices can be successful at promoting positive behavioural change as users are, fundamentally, unaware of their energy consumption (Attari et al., 2010). Consequently, individuals are unaware that they can, or indeed need, to take action to modify their behaviour. Previous research (Abrahamse et al., 2007) has demonstrated that household energy use can be significantly reduced following targeted interventions and advice that directly accounts for specific user behaviours. This approach has also been documented to be successful within previous work within the automotive sector, primarily those targeting the uptake of eco-driving behaviours (Andrieu & Saint Pierre, 2012; Tulusan et al., 2012; Mensing et al., 2014). These studies have uniformly identified that significant fuel savings are possible following the provision of in-vehicle feedback devices that respond to driver actions.

DESIGNING INTERFACES

The development of interfaces to encourage environmentally conscious behaviour can be seen as placing the designer as a controller of human behaviour. Whilst this role could be seen as beyond designers' remit, the design of objects has always had an irrefutable fundamental influence on subsequent activities (Simon, 1969; Redström, 2006). Whether the subject of the design is a desired physical object or an interface developed to direct or modify user behaviours, designers have an explicit role in influencing the decision making process (Lockton et al., 2009). This approach is perhaps best popularised by Fabricant's (2009) phrasing that *"Designers are in the behaviour business"* (cited in Lockton et al., 2010). The search for novel approaches to design is of growing interest to researchers (Schweitzer et al., 2014; Hatcher et al., 2018). One design approach that may be of value in this pursuit is the "Design with Intent" (DwI) toolkit (Lockton et al., 2010). From a foundation within ecological psychology (Barker, 1968), DwI seeks to combine an understanding of human activities with affordance theory (Gibson, 1986) with insights gained from prominent design theorists (Norman, 1999; McGrenere & Ho, 2000) to offer a flexible approach to novel design. The approach is predicated on the view that behaviour can be directed by design (Norman, 1999), with design having an intrinsic role in suggesting and promoting desirable behaviours whilst simultaneously constraining and reducing the potential for undesirable behaviours to occur. DwI acts as a *"Suggestion tool"* (Lockton et al., 2010), which seeks to inspire designers to develop novel solutions to problems.

The DwI approach is characterised by the use of 101 design cards, divided between 8 key lenses, each of which loosely corresponds to the theme of the cards. Many cards could fit into multiple lenses and the division between such lenses can often be seen as somewhat arbitrary (Buchanan, 1985). The key lenses are Architectural, Error-proofing, Interaction, Ludic, Perceptual, Cognitive, Machiavellian, and Security. Table 4.1 presents a summary of the main themes of each of the lenses, as well as example cards from the toolkit lens. Each DwI card presents a single question designers and developers can ask about their target product, system, or interface and a real-world example of that question in practice to act as an inspiration

TABLE 4.1
Lenses and Themes Presented within the DwI Framework

Lens	Theme	Example Card
Architectural Lens	Draws primarily on ideas within architecture and urban planning, seeking to apply ideas from the built environment. Concerns the structure and layout of items and behaviour.	Angles Pave the Cowpaths
Error-proofing Lens	Considers any behaviour that deviates from a target behaviour as an error and seeks to reduce the likelihood of errors occurring. Seeks to design a system whereby these errors cannot occur.	Are You Sure? Matched Affordances
Interaction Lens	Fundamentally about users' interaction with the devices or displays. Based on the feedback and feedforward of information between the user and the device being considered.	Kairos Real-Time Feedback
Ludic Lens	Focus on the potential for gamification of a device. Popularised by the view that playful interactions can encourage the maintenance of behavior.	Scores Storytelling
Perceptual Lens	Seeks to utilise biases in human perceptual system, for example use of heuristics, to target the design and development of objects.	Colour Associations Nakedness
Cognitive Lens	Based on cognitive psychology and an understanding of how individuals make decisions. Seeks to bias individuals to make a desired decision.	Provoke Empathy Commitment and Consistency
Machiavellian Lens	Seeks to control the behaviour of individuals, by utilising an "Ends Justify the Means" approach.	Functional Obsolescence I Cut, You Choose
Security Lens	Seek to prevent undesired behaviour through direct countermeasures. Seeks to directly control behaviour.	Peervailence Coercive Atmospherics

to help designers see potential applications of the card. Designers are required to use the information presented on the card to make their own inferences about their products and their end-users needs, with no pre-existing boundaries set in place. A key advantage of the DwI approach is that it is a simple approach which allows non-experts to design new products quickly and efficiently. As an example, whilst design approaches such as Design Sprint (Banfield et al., 2015) take five days to complete, DwI takes a single session to produce usable and innovative designs (Lockton, 2010).

Despite the freedom that DwI offers as a design tool, it could be argued that this approach lacks guidance on how to best structure ideas. Indeed, designers are never required to actively consider the fundamental requirements of the system or interface being developed nor consider end-users' needs, subsequently meaning that it is not possible to validate the generated ideas without significant further testing. To address this shortfall, the researchers considered whether established Human Factors methods aimed at developing and mapping the requirements of systems could be of benefit to users of the DwI toolkit, or act as a way to validate the subsequently produced designs. One such approach, popular within academic literature, is Cognitive Work Analysis (CWA) (Rasmussen, 1986; Stanton et al., 2018).

COGNITIVE WORK ANALYSIS

As seen in Chapter 2, CWA can act as an ideal starting point for the development of novel systems (Rasmussen, 1990). As a recap, CWA was originally developed for use in the nuclear power industry (Rasmussen, 1986) and acts as a structured framework for understanding complex socio-technical systems, systems in which people and technology are closely coupled (Stanton et al., 2018). Fuel-efficient driving is a suitable task for this analysis as drivers must interact with in-built vehicle mechanical systems, other road users, and increasingly, in-vehicle technology, including driver assist technology, and since the release of Tesla Model S in 2014, fully automated driving systems. CWA seeks to map the constraints that structure the working system, allowing practitioners to understand what is required of the system as well as what is both possible and not possible within the confines of system operations. By focussing on the constraints that frame a system, the analysis seeks to understand and support user needs for improved efficiency and safety. Drawing upon foundations in ecological psychology, general systems thinking and adaptive control systems (Fidel & Pejtersen, 2004), CWA has developed into a domain agnostic and highly flexible method that can be used to understand a variety of disciplines and also explore the potential of future system developments. CWA is an ideal method for envisioning revolutionary design as it promotes a focus on the fundamental requirements of the system (Naikar & Lintern, 2002). Due to related theoretical underpinnings, it is proposed within this chapter that the insights gained from CWA can be extended by DwI in order to develop usable interfaces with which users can directly interact. By combining the free flow idea generation of DwI with the constraint-based framework of CWA, designers are free to be creative within their designs, provided that the fundamental needs of the system are met. Tools to extend the CWA approach are needed as no typical means of using the outputs of CWA within design processes currently exist (Read et al., 2015).

As outlined in Chapter 2, developing a complete CWA is an extensive and time-consuming process, and the steps required to complete this analysis and the insights gained are presented there. The focus of this chapter is the initial idea generation following use of the DwI toolkit, extending the previously outlined CWA. There is, however, value in recapping some of the key aspects of CWA within the current chapter to allow for clearer understanding. The complete CWA process comprises of five key phases, Work Domain Analysis (WDA), Control Task Analysis (ConTA), Strategies Analysis (StrA), Social Organisation and Cooperation Analysis (SOCA), and Worker Competencies Analysis (WCA) (Vicente, 1999; McIlroy & Stanton, 2011). The key phase of the CWA considered within this chapter is the WDA. The primary focus of the WDA is the development of an abstraction hierarchy. The abstraction hierarchy aims to map the proposed system on multiple conceptual levels, ranging from its reason for existing to the physical objects that the system is comprised of. Five Conceptual levels are considered when developing an abstraction hierarchy. The uppermost level maps the system's "Functional Purpose(s)", the system's raison d'etre, or reason(s) to exist. Below this level, the system's "Values and Priorities" are presented. The "Values and Priorities" level maps metrics for measuring the system's success, how users and observers can know that their system is achieving the outlined "Functional Purpose(s)". The central level of the abstraction hierarchy is the "Purpose Related Functions". Within this level are functions linking the system's activity to the roles offered by each of its constituent components. The fourth level is "Object Related Processes". Within this level, the input of each "Physical Object" within the system is considered in terms of what it contributes to wider system functioning. The final, or foundation level of the abstraction hierarchy is the "Physical Objects" level, which documents all of the tangible objects of which the system is comprised. The generated abstraction hierarchy can be validated using an exhaustive means-ends analysis, following the why-what-how triad approach (McIlroy & Stanton, 2011). It is possible to nominate any item within the hierarchy and ask the question, "what does this do?" By examining all connections in the layer immediately above the node, it is possible to answer the question, "why does it do this?" When considering all connections in the layer immediately below, it must be possible to answer the question, "how does it achieve this?" This validation process ensures that all connections are suitable. Once completed, the abstraction hierarchy actively maps out the system for designers. This stage is considered essential for development as it can be seen as laying the foundation for the system under investigation. The abstraction hierarchy identifies the constraints on workers' behaviour based upon their physical context (Naikar, 2006). Given the focus of the current book, this can, unsurprisingly, be considered in terms of how the wider road environment, including both infrastructure and other road users, and the current vehicle context, including its technological capacities, influence the achievement of greater fuel efficiency.

CWA offers analysts a technology agnostic approach to consider a system, allowing for the consideration of both technology and human agents in the same analysis. This makes it an ideal approach for the consideration of novel technology as well as a tool to consider the constraints for a new interface in a previously established working environment. Despite these benefits, the final outcome of the CWA analysis is not a complete workable design of the envisaged system or interface. It is in this gap that this chapter is focussed, exploring the use of DwI to progress thinking towards initial mock-up designs, in preparation for further work empirically assessing the impact that such interfaces can have.

RESEARCH GOAL

In summary, this chapter will document the process of combining knowledge gained from a developed CWA documenting fuel-efficient driving with DwI in the design of in-vehicle interfaces. This is applied to two in-vehicle interface development case studies. This chapter will focus on the application of knowledge gained from the CWA to act as theoretical underpinning for interfaces developed using the DwI toolkit to examine the extent to which these methods can complement one another.

METHOD

Participants

To develop the interfaces, two main workshop sessions were held. The first workshop was comprised of two female participants, aged 26 and 39 years ($M = 32.5$), and one male participant, aged 24 years. All participants possessed a background in Human Factors and driving research, but did not have an understanding of fuel efficiency. The second workshop was comprised of three participants, two male participants, aged 32 and 33 years ($M = 32.5$), and one female participant, aged 31 years. Two of the participants held substantive backgrounds in Human Factors research. The third participant had considerable experience in the development of information displays, primarily for use by rail passengers. All participants were recruited via opportunity sampling and the use of a recruitment mailing list. Two of the three participants in each of workshop held a full UK driving license and had extensive experience driving on the UK road network. Participants were required to provide full informed consent prior to the start of the study. Although these groups are small, especially in line with work suggesting that innovation is positively correlated with group size (Hülsheger et al., 2009), practicalities of the study and participant availability restricted the use of larger samples. As the focus of the current work is to examine whether CWA and DWI could be integrated, two workshops were deemed preferable to a single case study workshop. As an additional advantage, smaller group sizes were beneficial in allowing a single research facilitator to better manage the workshops.

Procedure

The University of Southampton Ethics Committee gave full ethical approval for this study prior to the start of the workshops. Both workshops followed the same structure, however, due to differences in participants' backgrounds, experience, and the volume of discussion, timings varied between groups. Participants were initially introduced to the research program, the overall aims of the session, and received a brief introduction on the concept of eco-driving and improving fuel efficiency when driving through the modification of driver behaviour. Following this introduction, participants were introduced to the previously completed CWA documenting fuel-efficient driving (Stanton & Allison, 2020; Chapter 2). The previously completed CWA had mapped the potential constraints that would operate around an eco-driving interface, across multiple driver skill levels and driving scenarios, for example waiting at traffic lights and accelerating to higher speed. Participants had access to all elements of the completed CWA, including a large poster scale print of the abstraction hierarchy. Introduction to the project, eco-driving, and familiarisation with the CWA lasted approximately 45 minutes. During both introduction and familiarisation stages, participants were encouraged to ask questions to the research team about the wider topics of fuel efficiency and eco-driving as well as the CWA in order to encourage deeper consideration and understanding of the subject area.

Once participants had been familiarised with the project objectives and the previously completed CWA, they were presented with a single scenario. For Workshop 1 this scenario was waiting at traffic lights; for Workshop 2 the scenario considered was overtaking. A single specific scenario was chosen in order to better frame the workshops and make most use of the available time. Participants were asked, using the presented CWA, to design an interface that would help the driver to become more fuel-efficient during the presented scenario. Participants were asked to work through all 101 DwI cards (Lockton, 2010) whilst considering the scenario and the CWA (Stanton & Allison, 2020) to inspire suitable designs. Participants were informed that they were free to use any form of interaction display within their design, including head-down displays (HDDs), head-up displays (HUDs), auditory signals, and haptic signals. For the design element of the workshop, participants were presented with A3 sheets of paper, post-it notes, and a variety of different colored pens and actively encouraged to think in a creative manner when developing the required interfaces. Participants were asked to exhaustively consider whether each DwI card could, or should, be incorporated into the designed interface. Participants were told that they could either modify their existing design, or develop a new design incorporating their previous ideas with those generated by the use of further DwI cards. When participants introduced an interface element based upon a DwI card, a member of the research team asked them to discuss why and how this card informed their progressing design. The research team made substantive notes throughout this time to aid future understanding of the design. A member of the research team was on hand throughout the workshops to answer any questions that arose, to moderate the session, and to ensure that each participant was able to contribute ideas to the session. The research facilitator, however, did not attempt to influence the group designs in any way, and did not impose their opinions on the groups' designs during the workshops. Following the development of the initial design, the groups were asked to review their designs and ideas to ensure that all members of the group were happy to progress. Approximately 60 minutes was given to the design stage of the session, but participants were not explicitly timed.

The final phase of the workshop focussed on the use of the previously presented CWA (Stanton & Allison, 2020) to review and redesign the developed interface as appropriate. Participants were asked to reflect on all of the previously completed stages of the CWA, and discuss how each of the key elements within their interface was informed by the CWA. Despite this section of the workshop being largely a reflective exercise and a linear discussion process within the groups, it did spark considerable deliberation and discussion, lasting approximately 45 minutes. Participants were free to revisit their design and modify should they feel this was required. Following this stage, participants were offered the opportunity to reflect upon their use of the DwI cards and the overall workshop experience. Table 4.2 presents a summary of the different workshop phases and timings for clarity.

TABLE 4.2
Workshop Summary

Phase	Content	Timings	Input	Outcomes
Introduction and consent	Researcher outlines the current study, presenting participants with an information sheet and consent form.	5 Minutes		Participants aware of study design and requirements are able to give informed consent.
Introduction to fuel efficiency and familiarisation with the previously completed CWA	Participants are presented with an overview of fuel-efficient driving, the behavioural approaches to fuel usage, and familiarised with the completed CWA.	45 Minutes	Previously Completed CWA (Stanton & Allison, 2020)	Participants knowledge grounded within previously completed works and system operations.
Scenario presentation and interface design using DwI cards	Participants were presented with a scenario and asked to exhaustively use the DwI cards to design a suitable in-vehicle interface to support drivers in completing the scenario as fuel efficiently as possible.	60 Minutes	101 DwI cards (Lockton, 2010)	Initial interface(s) developed prior to further refinement.
Review and redesign of the developed interface using the previously presented CWA	Participants were asked to reflect on their completed interface(s) and discuss how each element was informed by the previously developed CWA. Elements which could not be explained using the previously developed CWA were refined or removed.	45 Minutes	Initial interface design & the previously completed CWA (Stanton & Allison, 2020)	Final developed interface(s).
Reflection	Participants were asked to reflect on their use of CWA, the DwI cards, and the workshop experience.	5 Minutes		Knowledge of participants experience using the methodology.

RESULTS AND DISCUSSION

Two interface mock-ups were developed from the workshops, following participants' designs. The interfaces presented here are initial mock-ups and presentation of ideas, and are not currently deployed in vehicles or simulators for testing.

WORKSHOP 1 – WAITING AT TRAFFIC LIGHTS

Design of the Display

The interface mock-up designed for the task of "Waiting at Traffic Lights" is presented within Figure 4.1. This scenario was chosen as it is a point in the drive where the driver is able to review their current performance without becoming distracted from the overall driving task and risking their safety. The interface devised was based on 47 unique DwI cards, across all 8 lenses. It should be noted that the interface was designed for future use, as it does account for the potential of interconnected vehicles and infrastructure, a potential explicitly presented within the CWA that participants used to guide their design.

The developed interface contains eight key elements, a countdown traffic light display, a potential to proceed display, a surround vision system, a fuel efficiency feedback display, a minimised satellite navigation display, a route selection display, a fuel gauge, and a radio/entertainment display. A summary of each interface element and the role each element fulfils is are provided in Table 4.3.

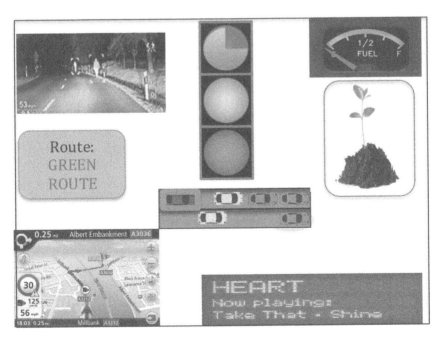

FIGURE 4.1 Designed HDD interface mock-up for the scenario "Waiting at traffic lights".

TABLE 4.3
Summary of Interface Elements

Element	Element Name	Element Explanation
1.	Count Down Traffic Light Display	Display indicating to drivers the approximate time the traffic lights will remain on their current color, allowing drivers to gauge relative wait time. In the current example, the lights are on red and have approximately a quarter of the time remaining.
2.	Potential to Proceed Display	Display indicating the likeliness a driver is to proceed through the next revolution of lights. Green colored cars indicate that the vehicle will proceed, yellow indicating it is possible the vehicle will proceed through the lights should the vehicles preceding it react to the changing traffic lights in a positive manner and red indicating that the vehicle is highly unlikely to pass through on the current revolution of the lights. The driver's vehicle is indicated by larger selection box. In the current example, this display is indicating that the driver is unlikely to proceed through the next revolution of traffic lights.
3.	Surround Vision System	Opportunity to provide drivers with a view of events happening to the sides and rear of the vehicle to increase the chance of detecting pedestrians who may be about to cross the road regardless of the indication on the traffic lights.
4.	Fuel Efficiency Feedback Display	The fuel efficiency presents a pictorial representation of the driver's current fuel efficiency based on their ability to follow eco-driving principles, such as gentle acceleration and braking. It is currently envisioned that this display will be customisable by the user to account for environmental motivations, using images of a growing tree, such as in the current example, or images of coins for individuals who are financially motivated.
5.	Minimised Satellite Navigation Display	Display drivers' current road position and allow drivers to potentially modify their route or destination whilst the vehicle is stationary.
6.	Route Selection Display	Opportunity for drivers to select a route based on fuel efficiency. Within the current example, the driver has selected the most fuel-efficient "Green" route.
7.	Fuel Gauge	The fuel gauge presents the vehicle's current remaining fuel level. It is advised that the driver in the example refuels soon.
8.	Radio/Entertainment Display	The interface also displays the driver's current entertainment system information, as available on traditional in-vehicle entertainment displays. As the vehicle is stationary within the given scenario it is possible to interact with the entrainment system with minimal safety implications.

Taming Design with Intent Using Cognitive Work Analysis

Each element within the interface was developed using the combined DwI and CWA approach as outlined previously. To provide an illustration of the use of the DwI cards, Table 4.4 provides a summary of the different DwI cards that inspired the design of the fuel efficiency feedback display. Also included within this table is the use of cards that were seen as generic and an inspiration for the wider display rather than any single element.

TABLE 4.4
DwI Cards Used to Inspire Design of the Fuel Efficiency Display within the "Waiting at Traffic Lights" Interface

Lens	Card	Description/Reasoning
Architectural	Converging and Diverging	Offer a fuel efficiency score to encourage engagement with the task of becoming fuel-efficient.
Architectural	Positioning	Only activate key sections of the display when stationary.
Architectural	Segmentation and Spacing	Divide the interface display into individual elements so that individuals can interact with individual elements.
Architectural	Simplicity	Use of pictorial representations wherever possible to encourage a simple and accessible display.
Error-proofing	Defaults	Default the display options to be the most fuel-efficient possible and focus on environmental rather than monetary gains from the system.
Error-proofing	Portions	Divide the interface into smaller elements and offer users different feedback for different achievements and actions.
Interaction	Kairos	Switch to a traffic light information display as the vehicle approaches the traffic light.
Interaction	Partial Completion	Show users their achievements so far, how much fuel they have saved in the current journey by being fuel-efficient.
Interaction	Progress Bar	Digital display/pictorial representation of a plant or pile of coins acts as a progress bar towards overall fuel efficiency goal.
Interaction	Real-Time Feedback	Digital display/pictorial representation of a plant or pile of coins acts as real-time feedback to fuel usage and potential emissions.
Interaction	Summary Feedback	Give information about current performance via pictorial representation.
Interaction	Tailoring	Offer option to change plant representation to financial information represented by a pile of coins.
Ludic	Challenges and Targets	Allow users to set their own personalised fuel efficiency goals to reach in order to gain achievements.

(Continued)

TABLE 4.4 (Continued)
DwI Cards Used to Inspire Design of the Fuel Efficiency Display within the "Waiting at Traffic Lights" Interface

Lens	Card	Description/Reasoning
Ludic	Collections	Allow permanent collection of achieved goals/add the option to grow a permanent "garden".
Ludic	Levels	Achieve rewards at staggered levels of achievement on the pictorial representations to encourage greater engagement with the task of fuel-efficient driving as the journey continues.
Ludic	Rewards	Potential to gain visual rewards and permanent achievements based on actions.
Ludic	Scores	Give comparative behaviour feedback to encourage future behaviour, so that a driver must improve their fuel efficiency in order to gain the same level of reward.
Perceptual	Metaphors	Use of pictorial representation to make fuel saving more apparent to the driver.
Cognitive	Assuaging Guilt	Visual representation of a plant growing to encourage guilt reduction.
Cognitive	Commitment and Consistency	Encourage users to buy into the overall idea of reducing carbon footprint by incorporating environmental or financial ideas into the display.
Cognitive	Emotional Engagement	Encourage users to engage with the idea that fuel saving is the correct thing to do for both the environment and their financial wellbeing.
Cognitive	Habits	No significant changes in driver's current actions are required. The interface acts as an information prompt.
Cognitive	Rephrasing and Renaming	Potential to reframe eco-driving and emissions saving to a direct financial saving.
Machiavellian	Bundling	Pairing fuel saving or financial saving with emission reduction so that in order to save money the user consequently reduces emissions.
Machiavellian	Worry Resolution	Reduce worry caused by anti-environmental action of driving by displaying positive environmental images when driver is fuel-efficient.

Validation of the Display

In order to ensure that the display adhered to the previously completed CWA (Stanton & Allison, 2020), each element of the interface was compared against this documentation. Due to its focus on mapping the physical objects that comprise a system as well as the overall aims and objectives of the system, the abstraction hierarchy created as part of the WDA component was seen as the primary validation tool.

When considering the abstraction hierarchy and taking the example of the fuel efficiency display (Item 4, Table 4.3), this item can be linked to the functional purposes of "Save Energy" and "Reduce Emissions", holding the values and priorities of "Optimise Vehicle Range", "Reduce Fuel Usage", "Optimise Driver Satisfaction", "Reduce NOx", and "Reduce CO_2". It does this by accounting for and providing drivers more information regarding the Purpose Related Function "Control Vehicle Motion". To calculate the relative success of the driver and be able to contribute feedback, the display is able to present information to the driver related to their ability to "Control Acceleration" and "Control Vehicle Speed", have knowledge of the "Speed Limit" and encourage "Smooth Motion". In order to achieve these goals, the device can take information from the vehicle, as captured within the Physical Objects including "Clutch", "Fuel", "Brake Pedal", and "Accelerator Pedal". In addition, this application is reliant on the physical object "V2X Communication" to allow it to accurately communicate with surrounding infrastructure to allow presentation of the lights duration and offer an estimation of approximate dwell time. The corresponding nodes from the abstraction hierarchy are presented in Figure 4.2, mapping how this display element can be used to reduce emissions.

A similar validation process was undertaken for all interface elements in order to ensure that the functioning of each element was warranted based upon the previously generated specifications. Using the generated CWA as a validation tool helps to ensure that each interface element can contribute to the primary function

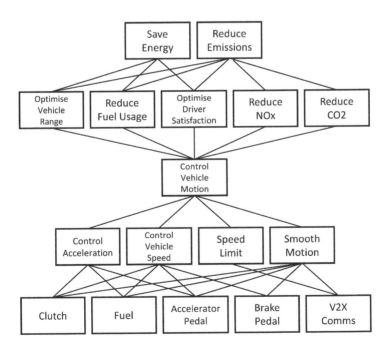

FIGURE 4.2 Subset of the Abstraction Hierarchy accounted for by the fuel efficiency display (Item 4, Table 4.3).

of the system. In this way the developed interface can be seen to support users in achieving greater fuel efficiency. Of note is that this interface display makes use of both feedback systems, such as shown by Fuel Efficiency Feedback Display (Table 4.3, Item 4), but also feedforward information, provided by the Count Down Traffic Light Display (Table 4.3, Item 1) and Potential to Proceed Display (Table 4.3, Item 2). By providing feedback on behaviour, it is hoped that long term drivers develop positive driving habits. By providing feedforward information, drivers will be aware of both the time they have to wait, removing any need for anticipatory actions, and the likelihood of passing through the lights allowing for more gentle acceleration within the traffic flow if they would be required to again wait at the lights.

WORKSHOP 2 – ACCELERATING TO OVERTAKE

Design of the Display

The interface designed for the task of "Accelerating to Overtake" is presented within Figures 4.3 and 4.4. This scenario of overtaking is not associated with the typical activity of fuel-efficient driving; however, it is an activity that many drivers are likely to engage in on a regular basis. The devised interface was based on 29 unique DwI cards and, similar to the previous "Waiting at Traffic Lights" interface, utilised all 8 lenses. Unlike the previous interface display, which only used the vehicle's HDD, the overtaking interface is primarily presented as part of the vehicle's HUD (Figure 4.3) in order to remove the need for the driver to divert their gaze away from the road ahead. This information can be supplemented with auditory feedforward information

FIGURE 4.3 Designed HUD interface mock-up for the scenario "Accelerating to overtake". Within the current image the vehicle has overtaken the vehicle in the middle lane and is being informed they should be prepared to follow the ghost car in moving back to the middle lane.

Taming Design with Intent Using Cognitive Work Analysis

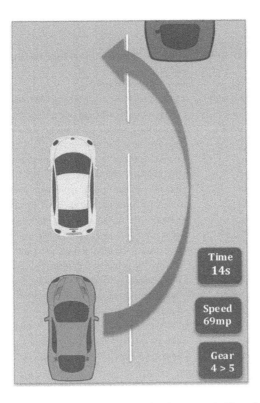

FIGURE 4.4 Designed HDD interface mock-up for the scenario "Accelerating to overtake". Within the HDD interface the driver, presented in the rear green car is being informed that in 14 seconds it will be safe to overtake the preceding yellow car, and is informed that the optimal overtaking speed is 69 mph in 5th gear.

presented to the driver prior to the start of the manoeuvre. A breakdown of the task and the details of the actions that will be undertaken, supporting this auditory feedforward information is presented in the vehicle's HDD for redundancy, as shown in Figure 4.4.

When considering the "Accelerating to Overtake" HUD display (Figure 4.3), the key novel feature is the presentation of the ghost car. The ghost car presents users with an ideal model of how to complete the task of overtaking, offering guidance on timing and speed, constrained with the view of completing the manoeuvre in the most fuel-efficient way possible. This idea has been heavily influenced by the use of ghost cars that are popular in gaming. The HDD display, in contrast, is not designed to present any novel information to the driver, but rather reinforce information that the driver may have missed from the audio system, including time until manoeuvre and the ideal speed the car should travel in order to complete the manoeuvre and remain fuel-efficient. Table 4.5 provides a summary of the use of different DwI cards that inspired design of the fuel efficiency feedback display, it includes both HUD and HDD elements.

TABLE 4.5
DwI Cards Used to Inspire Design of the "Accelerating to Overtake" Interface

Lens	Card	Description/Reasoning
Architecture	Converging and Diverging	Channel people into different lanes so they safely split up and allow room for efficient overtaking.
Architecture	Conveyor Belts	Overtaking interface only appears following request – button press or similar interaction.
Architecture	Mazes	Encourage following of the most fuel-efficient path when overtaking to still achieve goal but in a fuel-efficient way.
Architecture	Positioning	Only allow access to the application when appropriate and safe.
Architecture	Roadblock	Adjust the speed or limit speed of the car by use of a ghost car.
Architecture	Simplicity	Design with simplicity and use of HUD rather than HDD.
Error-proofing	Are You Sure?	Ghost car only appears as car starts manoeuvre and/or is requested.
Error-proofing	Choice Editing	Only present the desired route and ghost car information to remove the potential for choice.
Error-proofing	Defaults	Limit available guidance to be the best for the current road situation.
Error-proofing	Did You Mean?	Visual indication to driver if they are not at the ideal speed for current road situation.
Interaction	Kairos	Inform users of the best moment to make an overtaking manoeuvre.
Interaction	Progress Bar	Show on HUD progress compared to a ghost car.
Interaction	Tunnelling and Wizards	Ghost car acts as a wizard to guide users.
Ludic	Challenges and Targets	User is challenged to match as closely as possible to ghost car to achieve greatest fuel efficiency.
Perceptual	(A)Symmetry	Match HUD and change actions to show clear link.
Perceptual	Colour Association	Colour changes on car in interface for when car considering an overtaking manoeuvre Red = Bad/Not now, Green = Good/Go now.
Perceptual	Contrast	Flashing box on the display surrounding car to stress good speed or Revolutions Per Minute (RPM) required to complete the manoeuvre in a fuel-efficient way.
Perceptual	Implied Sequences	Ghost car changes and progresses in manoeuver following driver actions.
Perceptual	Mood	Use of Green = Good to encourage link between actions and environmental impact.
Perceptual	Perceived Affordances	Overtaking lines and guidance only appear when it is safe to perform an overtaking action.
Perceptual	Prominence	Use of HUD display to make ghost car salient and obvious to the driver.
Perceptual	Seductive Atmospherics	Use of ambient sounds to match vehicles RPM and make eco-driving more immediate to driver.
Cognitive	Commitment and Consistency	Commit users to the idea that eco-driving does not relate to a slow or frustrating journey. And be consistent in display and information provision to help easier buy-in.
Cognitive	Expert Choice	Ghost car acts as an expert "user" to suggest the best action to take in the current situation.

(Continued)

Taming Design with Intent Using Cognitive Work Analysis

TABLE 4.5 (Continued)
DwI Cards Used to Inspire Design of the "Accelerating to Overtake" Interface

Lens	Card	Description/Reasoning
Cognitive	Habits	Make it easy for people to use the system and make use of the overtaking application so that it becomes ingrained into actions.
Machiavellian	Forced Dichotomy	No middle ground available in the system, only expert choices are presented when the system is activated.
Machiavellian	Format Lock-in	Eco-based system is the only option, if users want overtaking advice, they see the most fuel-efficient way of achieving this goal.
Machiavellian	Serving Suggestions	Display ideal speed and RPM to the driver before starting the overtaking manoeuvre.
Security	Where You Are	Disable the option for the overtaking app to work on roads where overtaking is not an option.

Validation of the Display

Similar to the "Waiting at Traffic Lights" interface, the "Accelerating to Overtake" interface was compared to the CWA abstraction hierarchy, and appropriate elements were identified as shown in Figure 4.5. To ensure that the interface fulfilled functional purposes of "Save Energy" and "Reduce Emissions (CO_2 and NOx)". Within this interface these goals are achieved as they adhere to the Values and Priorities of "Minimise Traffic Delay", "Minimise Congestion", "Optimise Driver Satisfaction", and "Optimise Travel Time". Whilst overtaking is generally not considered fuel efficient, due to the additional fuel required to accelerate the vehicle, overtaking may "Optimise Vehicle Range" and "Reduce Fuel Usage" if the driver is able to shift to a higher gear. Assuming a form of combustion engine, these engines are more efficient

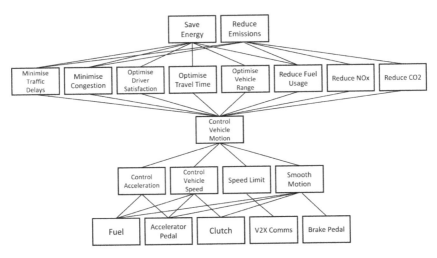

FIGURE 4.5 Subset of the Abstraction Hierarchy accounted for by the "Accelerating to Overtake" display.

at higher gears, potentially allowing the Values and Priorities of "Reduce NOx" and "Reduce CO_2," to be achieved. The system achieves these goals by accounting for and providing drivers more information regarding the Purpose Related Function "Control Vehicle Motion". To calculate the relative success of the driver and be able to provide feedforward information to the driver, the display presents information to the driver related to their ability to "Control Acceleration" and "Control Vehicle Speed", have knowledge of the "Speed Limit" and encourages "Smooth Motion". In order to achieve these goals, the device can take information from the vehicle, as captured within the Physical Objects including "Clutch", "Fuel", "Brake Pedal", and "Accelerator Pedal". In addition, this application is reliant on the physical object "V2X Communication" to allow the vehicle to accurately communicate with nearby vehicles and infrastructure to allow accurate estimation of the moment the driver can safely overtake.

GENERAL DISCUSSION

The aim of this chapter was to present initial work examining use of CWA (Rasmussen, 1986) to constrain interfaces developed using DwI (Lockton et al., 2010) to encourage fuel-efficient driving. By focussing on user requirements, as provided by CWA (Stanton & Allison, 2020), it was seen that the discussions relating to the DwI cards were highly structured and directed. By providing the CWA and DwI cards, individuals without a background in design or fuel-efficient driving were able to develop initial mock-ups of potential interfaces. Previous research (Read et al., 2015) has highlighted the need to extend the insights offered by CWA and the present investigation suggests that DwI is an appropriate tool for this goal.

DwI was envisioned as a *"suggestion tool"* (Lockton et al., 2010) to inspire novel designs. Within the current study, DwI was used to develop interfaces that aim to reduce fuel use and emissions whilst driving to limit the negative impact such emissions can have, both on human respiratory health (McCubbin & Delucchi, 1999; Buckeridge et al., 2002) and the wider eco-system (Karl & Trenberth, 2003; Chapman, 2007). Although neither interface directly interacted with the driver or the vehicular controls, the presence of such interfaces may be sufficient to encourage greater fuel efficiency as the drivers become more aware of their overall fuel use (Abrahamse et al., 2007). By actively promoting fuel efficiency within everyday driving, drivers can become aware of the impact that their behaviour can have and consequently take steps to reduce both their fuel usage and corresponding emissions (Stanton & Allison, 2020).

CWA seeks to exhaustively map a domain in order to facilitate extensive understanding and allow informed decisions to be made in response to system redevelopment (Rasmussen, 1986). However, CWA lacks a clear avenue for progressing insights into workable interfaces. In contrast, DwI (Lockton et al., 2010) is a toolkit to aid novel design, but lacks any true grounding to ensure that the designs developed meet user needs and requirements. Within the current chapter, it is argued that CWA can be used to inform, guide, and constrain the generated interfaces developed using DwI. In turn, it was found that CWA could be used to validate the proposed interfaces developed using DwI, ensuring that the DwI cards were able to actively address the fundamental requirements of the system under investigation. Although this chapter only provides a case study combining the methods, it is hoped that future research can develop a formalised approach to provide best practice guidance on using both CWA and DwI to develop Human-System interfaces. Whilst previous research suggests that there are no common means of using the outputs of CWA within design presently (Read et al., 2015), the current chapter provides a clear avenue regarding using the CWA methodology within interface design.

It is clear that the interface mock-ups presented in Figures 4.1, 4.3, and 4.4 are not ready for immediate deployment in a vehicle or simulator testing facility and require further work in order to make sure they are aesthetically pleasing. This research has not directly considered the importance of aesthetics in interface design and development, and the created interface mock-ups would benefit from input from a designer to improve visual appeal. Provided that all features of the display are maintained, developers can be sure that the interface fulfils user needs and requirements for the goal of minimising fuel consumption. Therefore, it is important that the combination of CWA and DwI happens early in an interface development cycle. Extending this point, initial interface design is but the first step in the design journey (Nielsen, 1993; Allison & Stanton, 2020). Testing is required, both in laboratory and field studies, to fully appreciate end-users' engagement in the displays.

It should be noted that a limitation of the chapter is that participants were only presented with the combined CWA and DwI cards, so it is not possible to assess the direct influence that either of these elements held over the final designs, or indeed whether the designs generated within the research would be substantially different were they developed by a different team which lacked these resources. This research was intended to look at the potential for the combination of the CWA and DwI approaches and considerably more work is required to elucidate the relative value in this approach.

Future research is needed to examine the extent to which a constrained DwI approach can develop novel ideas for deployment in vehicles. It would also be useful to present the developed ideas to a variety of external potential end-users in order to gain feedback and provide practical validation beyond that gathered from the theoretical validation offered by CWA. This further validation will enable researchers to identify ideas worthy of further pursuit, including the potential to explore the impact of both interfaces within an empirical, user-focussed, simulator study, whereby fuel savings and overall interface effectiveness can be directly assessed.

CONCLUSIONS

This chapter has presented a proof of concept that the open and domain agnostic toolkit, DwI, could be constrained and used to develop interfaces when supported by the Human Factors method, CWA. Participants, individuals without a background in fuel-efficient driving or design, were able to take the insights gained from the CWA process and confidently work through the DwI toolkit to develop potential interfaces. It can be argued that the DwI toolkit allowed participants to create initial concepts, whilst CWA acted to constrain the ideas to ensure that they remained focussed on the end goals of the system. This study acts as a proof-of-concept that combining these two distinct methodologies is possible and, more importantly, offers a potentially valuable approach when developing interface concepts that are grounded within design principles.

REFERENCES

Abrahamse, W., Steg, L., Vlek, C., & Rothengatter, T. (2007). The effect of tailored information, goal setting, and tailored feedback on household energy use, energy-related behaviors, and behavioral antecedents. *Journal of Environmental Psychology*, 27(4), 265–276.

Allison, C. K., & Stanton, N. A. (2019). Eco-driving: The role of feedback in reducing emissions from everyday driving behaviours. *Theoretical Issues in Ergonomics Science*, 20(2), 85–104.

Allison, C. K., & Stanton, N. A. (2020). Constraining design: Applying the insights of cognitive work analysis to the design of novel in-car interfaces to support eco-driving. *Automotive Innovation*, 3(1), 30–41.

Andrieu, C., & Saint Pierre, G. (2012). Comparing effects of eco-driving training and simple advices on driving behavior. *Procedia-Social and Behavioural Sciences*, 54, 211–220.

Attari, S. Z., DeKay, M. L., Davidson, C. I., & De Bruin, W. B. (2010). Public perceptions of energy consumption and savings. *Proceedings of the National Academy of Sciences*, 107(37), 16054–16059.

Banfield, R., Lombardo, C. T., & Wax, T. (2015). *Design sprint: A practical guidebook for building great digital products*. O'Reilly Media, Inc, Sebastopol, California.

Barkenbus, J. N. (2010). Eco-driving: An overlooked climate change initiative. *Energy Policy*, 38(2), 762–769.

Barker, R. G. (1968). *Ecological psychology: Concepts and methods for studying the environment of human behavior*. Stanford University Press, Redwood City, California.

Barth, M., & Boriboonsomsin, K. (2009). Energy and emissions impacts of a freeway-based dynamic eco-driving system. *Transportation Research Part D: Transport and Environment*, 14(6), 400–410.

Buchanan, R. (1985). Declaration by design: Rhetoric, argument, and demonstration in design practice. *Design Issues*, 2, 4–22.

Buckeridge, D. L., Glazier, R., Harvey, B. J., Escobar, M., Amrhein, C., & Frank, J. (2002). Effect of motor vehicle emissions on respiratory health in an urban area. *Environmental Health Perspectives*, 110(3), 293–300.

Chan, C. C. (2007). The state of the art of electric, hybrid, and fuel cell vehicles. *Proceedings of the IEEE*, 95(4), 704–718.

Chapman, L. (2007). Transport and climate change: A review. *Journal of Transport Geography*, 15(5), 354–367.

Department for Environment Food and Rural Affairs (2017). UK AIR Air Information Resource. Retrieved from https://uk-air.defra.gov.uk/air-pollution/causes.

Fidel, R., & Pejtersen, A. M. (2004). From information behaviour research to the design of information systems: The cognitive work analysis framework. *Information Research: An International Electronic Journal*, 10(1), paper 210.

Gibson, J. J. (1986). *The ecological approach to visual perception*. Hills-dale, NJ: Erlbaum.

Hatcher, G., Ion, W., Maclachlan, R., Marlow, M., Simpson, B., & Wodehouse, A. (2018). Evolving improvised ideation from humour constructs: A new method for collaborative divergence. *Creativity and Innovation Management*, 27(1), 91–101.

Hill, N., Brannigan, C., Smokers, R., Schroten, A., Van Essen, H., & Skinner, I. (2012). EU Transport GHG: Routes to 2050 ii. Final Project Report Funded by the European Commission's Directorate-General Climate Action, Brussels.

Hülsheger, U. R., Anderson, N., & Salgado, J. F. (2009). Team-level predictors of innovation at work: A comprehensive meta-analysis spanning three decades of research. *Journal of Applied Psychology, 94*(5), 1128.

Jain, R. K., Taylor, J. E., & Peschiera, G. (2012). Assessing eco-feedback interface usage and design to drive energy efficiency in buildings. *Energy and Build, 48*, 8–17.

Jamson, A. H., Hibberd, D. L., & Merat, N. (2015). Interface design considerations for an in-vehicle eco-driving assistance system. *Transportation Research Part C: Emerging Tech, 58*, 642–656.

Karl, T. R., & Trenberth, K. E. (2003). Modern global climate change. *Sci, 302*(5651), 1719–1723.

Li, L., Gong, Y., Deng, J., & Gong, X. (2018). CO_2 reduction request and future high-efficiency zero-emission argon power cycle engine. *Automotive Innovation, 1*(1), 43–53.

Lockton, D. (2010). Design with intent: 101 patterns for influencing behaviour through design. *Equifine*, Windsor

Lockton, D., Harrison, D. J., & Stanton, N. A. (2009). Choice architecture and design with intent. NDM9, 9th Bi-Annual International Conference on Naturalistic Decision Making (Doctoral Consortium Proceedings). London, 23–26 June.

Lockton, D., Harrison, D., & Stanton, N. A. (2010). The design with intent method: A design tool for influencing user behaviour. *Applied Ergonomics, 41*(3), 382–392.

Martin, E., Chan, N., & Shaheen, S. (2012). How public education on eco-driving can reduce both fuel use and greenhouse gas emissions. *Transportation Research Record: Journal of the Transportation Research Board*, 2287, 163–173.

McCubbin, D. R., & Delucchi, M. A. (1999). The health costs of motor-vehicle-related air pollution. *Journal of Transport Economics and Policy, 33*(3), 253–286.

McGrenere, J., & Ho, W. (2000, May). Affordances: Clarifying and evolving a concept. In: Proceedings of Graphics Interface, Montreal, 179–186.

McIlroy, R. C., & Stanton, N. A. (2011). Getting past first base: Going all the way with cognitive work analysis. *Applied Ergonomics, 42*(2), 358–370.

McIlroy, R. C., Stanton, N. A., & Harvey, C. (2013). Getting drivers to do the right thing: A review of the potential for safely reducing energy consumption through design. *IET Intelligent Transport Systems, 8*(4), 388–397.

Mensing, F., Bideaux, E., Trigui, R., Ribet, J., & Jeanneret, B. (2014). Eco-driving: An economic or ecologic driving style? *Transportation Research Part C: Emerging Technologies, 38*, 110–121.

Naikar, N. (2006). Beyond interface design: Further applications of cognitive work analysis. *International Journal of Industrial Ergonomics, 36*(5), 423–438.

Naikar, N., & Lintern, G. (2002). A review of "Cognitive work analysis: Towards safe, productive, and healthy computer-based work" by Kim J. Vicente. *The International Journal of Aviation Psychology, 12*(4), 391–400.

Nielsen, J. (1993). Iterative user-interface design. *Computer, 26*(11), 32–41.

Norman, D. A. (1999). Affordance, conventions, and design. *Interact, 6*(3), 38–43.

Rasmussen, J. (1986). Information Processing and Human-Machine Interaction. An Approach to Cognitive Engineering, New York, North Holland.

Rasmussen, J. (1990). The role of error in organizing behaviour. *Ergon, 33*(10–11), 1185–1199.

Read, G. J., Salmon, P. M., & Lenné, M. G. (2015). Cognitive work analysis and design: Current practice and future practitioner requirements. *Theoretical Issues in Ergonomics Science, 16*(2), 154–173.

Redström, J. (2006). Towards user design? On the shift from object to user as the subject of design. *Des. Stud, 27*(2), 123–139.

Schweitzer, F., Gassmann, O., & Rau, C. (2014). Lessons from ideation: Where does user involvement lead us? *Creativity and Innovation Management, 23*(2), 155–167.

Simon, H. A. (1969). *The sciences of the artificial.* Cambridge, MA: MIT Press.

Stanton, N. A., & Allison, C. K. (2020). Driving towards a greener future: An application of cognitive work analysis to promote fuel-efficient driving. *Cognition, Technology & Work, 22*(1), 125–142.

Stanton, N. A., Salmon, P. M., Walker, G. H., & Jenkins, D. P. (2018). *Cognitive work analysis applications, extensions and future directions*. Boca Raton, FL: CRC Press.

Stillwater, T., & Kurani, K. S. (2013). Drivers discuss eco-driving feedback: Goal setting, framing, and anchoring motivate new behaviors. *Transportation Research Part F: Traffic Psychology and Behaviour, 19*, 85–96.

Tulusan, J., Staake, T., & Fleisch, E. (2012, September). Providing eco-driving feedback to corporate car drivers: What impact does a smartphone application have on their fuel efficiency? In: *Proceedings of the 2012 ACM Conference on Ubiquitous Computing* (pp. 212–215). ACM.

Vicente, K. J. (1999). *Cognitive work analysis: Towards safe, productive, and healthy computer-based work*. Mahweh, NJ: LEA.

5 Applying Design with Intent to Support Creativity in Developing Vehicle Fuel Efficiency Interfaces

INTRODUCTION

This chapter continues to explore the application of the Design with Intent (DwI) toolkit (Lockton et al., 2010) for the creation of ideas for in-vehicle interfaces designed to support fuel-efficient driving behaviours. The process of using the DwI toolkit to generate novel ideas is discussed, with a focus on the creative process of ideation (Pannells & Claxton, 2008). This chapter seeks to expand on previous work that has sought to influence pro-social behaviours through the use of design (Cash et al., 2017; Cash et al., 2017). The desire to leverage pro-social and pro-environmental behaviour is a growing desire and concern within both the ergonomics community (Thatcher et al., 2013) and the design community (Tromp & Hekkert, 2016). Rather than focussing on the physical aspects that make up a design, researchers are increasingly considering the impact that design can have on an individual's behaviour (Wever et al., 2008). One significant global challenge faced in the first half of the 21st century, and requiring significant resources to address, is the impact of anthropometric climate change (Ramanathan & Feng, 2009). All disciplines have a role in exploring the steps that can be taken to limit the currently high levels of greenhouse gas emissions.

Design methods are tools that can be used to facilitate creative thinking (Tromp & Hekkert, 2016), helping users generate ideas and to structure thinking on particular topics (Daalhuizen, 2014). This chapter continues to address the use of one such design method, the DwI toolkit (Lockton et al., 2010) to inspire novel designs for in-vehicle interfaces intended to reduce the fuel use and emissions associated with everyday driving.

VEHICLE FUEL EFFICIENCY

Human actions have an undeniable impact on our environment and climate (Thornton & Covington, 2015). Barkenbus (2010) argues that approximately 8% of the world's total carbon dioxide (CO_2) emissions is a direct consequence of personal transportation. Within the European Union (EU), approximately 25% of total CO_2 emissions relate to transportation, the second highest sector after electricity generation (Hill et al., 2012). The United States follows a similar trend with approximately 32–41% of total emissions being related to personal transportation (Bin & Dowlatabadi, 2005; Vandenbergh & Steinemann, 2007). When considering the EU's transport emissions, 72% relate directly to road transportation, principally privately owned cars (Hill et al., 2012). Vehicle emissions including gasses such as CO_2, carbon monoxide (CO), and nitrous oxides (NOx) as well as particulate matter, adversely impact our environment (Ramanathan & Feng, 2009) and negatively impact human health and wellbeing, with exposure to vehicle emissions being associated with an increased risk of developing a variety of respiratory based disorders, including asthma, and bronchitis (Buckeridge et al., 2002). Due to the scale of transportation and automobile related emissions, relatively small-scale actions and savings repeated across this domain can produce substantive reductions in emissions.

Although considerable technical developments within the automotive industry has led to significant improvements in vehicle fuel efficiency, further progress is still needed (Lorf et al., 2013). Recent technical developments include the production of more energy efficient and cleaner drivetrains including the use of electric, hydrogen fuel cell, and hybrid vehicles (Chan, 2007). The use of such vehicles is associated with greater efficiency and lower emissions (Gardner & Stern, 2008). The initially high investment cost of new vehicles and, in some cases, a corresponding lack of required infrastructure (Philipsen et al., 2018), means that replacing all older vehicles with newer and more efficient vehicles is not a viable solution for addressing the current levels of vehicle pollutants, at least not in the short-term. Rather than replacing older vehicles with newer, more efficient models, Barkenbus (2010) advocated that altering the way a vehicle is driven could significantly reduce fuel use and vehicle emissions. Barkenbus argued that by adopting a refined driving style, typified by modest acceleration, early gear changes, limiting the engine to approximately 2,500 revolutions per minute (RPM), anticipating traffic flow to minimise braking, driving below the speed limit, and limiting unnecessary idling, drivers could dramatically cut their fuel usage and subsequent emissions. This driving style is commonly referred to as eco-driving (McIlroy et al., 2013), and is possible with both traditional internal combustion engines and newer hybrid vehicles (Franke et al., 2016). Offering empirical support for the use of these strategies, within a series of simulator trials, Birrell et al. (2013) found that the adoption of eco-driving behaviours resulted in significantly less fuel use without dramatically increasing journey time. Extending the work of Barkenbus (2010), Sivak and Schoettle (2012) suggest that the concept of eco-driving should also include strategic and tactical decisions. Strategic decisions relate to vehicle selection and ongoing vehicle maintenance, for example ensuring tyres are adequately inflated. Conversely, tactical decisions relate to routine decisions that drivers make on a daily basis, such as navigational decisions, for example changing route in order to avoid congestion and minimising the vehicles current load. By following these guidelines drivers can benefit financially, by using less fuel, and benefit the environment, by producing less greenhouse emissions and other pollutants without compromising safety (Young et al., 2011).

Whilst eco-driving can offer considerable benefits to both driver and the environment, previous work has shown that feedback and support are required in order to encourage the long-term adoption and maintenance of these behaviours (Tulusan et al., 2012; McIlroy et al., 2013; McIlroy & Stanton, 2017a; Allison & Stanton, 2019). This feedback and driver support can most easily be achieved by the use of in-vehicle interfaces. The design of interfaces has always had an irrefutable and fundamental influence on subsequent human activities (Simon, 1969; Redström, 2006). Due to this influence, designers can be seen as having an explicitly intended role in influencing decisions (Lockton et al., 2008). This approach is perhaps best popularised by Fabricant's (2009) phrasing that *"Designers are in the behaviour business"* (Cited in Lockton et al., 2016). Within this context, designers have a role to play in encouraging the adoption of fuel-efficient eco-driving behaviours. Acting as a form of persuasive technology, defined as *"a class of technologies that are intentionally designed to change a person's attitude or behaviour"* (IJsselsteijn et al., 2006), in-vehicle interfaces have the potential to change driver's relationship with their fuel efficiency and dramatically reduce their wider environmental impact (McIlroy et al., 2013; Allison & Stanton, 2020).

The design of novel interfaces to encourage a greater awareness of energy usage is not novel, and has been significantly pursued in previous research to reduce both household energy usage (Jain et al., 2012; Revell & Stanton, 2018) and in vehicle energy usage whilst driving (Jamson et al., 2013; McIlroy et al., 2017b). Attari et al. (2010) propose that such devices can be successful at changing behaviour as users are fundamentally unaware of their energy use and how their actions contribute to overall energy use, as a consequence, users are unaware of how they can modify their behaviour to be more energy efficient. Previous research has demonstrated that energy use within household environments can be significantly reduced following targeted interventions either fostering greater understanding (Revell & Stanton, 2014; Revell & Stanton, 2016) or advice accounting directly for users' specific behaviours (Abrahamse et al., 2005). This approach has also been documented to be a great success within previous research focussed specifically within the automotive sector, primarily those targeting the uptake of eco-driving behaviours (Tulusan et al., 2012; Mensing et al., 2014). These studies have uniformly identified significant fuel savings as a result of providing in-vehicle eco-driving feedback devices that respond directly to drivers' actions. With both a need and precedent for the use of in-vehicle interfaces to address fuel usage, work is needed to design suitable interfaces which are effective at reducing fuel use and emissions, whilst also being accepted by users.

DESIGN WITH INTENT TOOLKIT

One design approach to aid in the creative development of both physical items and behaviour is the "Design with Intent" toolkit (Lockton et al., 2010). The DwI toolkit can be viewed as a collection of design patterns that can be used to guide the development of novel systems and interfaces. From a fundamental perspective, DwI was heavily influenced by ecological psychology (Barker, 1968; Gibson, 1986), and stresses how individuals' actions can be directed and constrained, if not directly controlled, by their environment. The DwI approach is, therefore, predicated on the view that behaviour can be directed by design (Buchanan, 1985), both promoting potential desirable behaviours as well as constraining and reducing the potential for undesirable behaviours to occur. To encapsulate this approach, the DwI toolkit was heavily influenced by the work of Alexander (1977) within architecture and the work of Tidwell (2005) from a purer design perspective. By combining an understanding of human activities, informed by affordance theory (Gibson, 1986) with the work of established design theorists (Norman, 1999; McGrenere & Ho, 2000), DwI provides a flexible toolkit to allow designers and engineers to generate novel design ideas that can later be matured into full interfaces and systems. In this regard, DwI acts as a *"Suggestion tool"* (Lockton et al., 2010, *p383*), which seeks to inspire designers and engineers to develop novel solutions to problems. By its approach as a suggestion tool, the DwI method seeks to act as a source of inspiration, with the tools being documented to trigger innovative and creative design solutions (Eckert & Stacey, 2000).

The DwI approach is characterised by the use of 101 design cards, divided between 8 key lenses, each of which corresponds to the themes of the card. Lockton et al. (2010) argue that the intention of each of the lenses is to focus around a particular worldview of a problem, for example, a designer with experience in safety engineering will have a focus on error-proofing, seeking to prevent deviations from safe behaviour. In contrast, an architect will have a focus on the potential of layout and use of space within an interface, applying physical design metaphors to the digital environment. Whilst Lockton et al. (2010) note that many cards could fit into multiple lenses, they argue that the different lenses aim to offer alternative perspectives on a problem, and encourage individuals to think outside their immediate frame of reference. Although originally directly building on de Bono's (2017) "six thinking hats" approach and comprising of six lenses, Architectural, Error-proofing, Pervasive, Visual, Cognitive, and Security (Lockton et al., 2010), the DwI approach has, through multiple iterations and testing (Lockton et al., 2008; Lockton et al., 2010; Lockton et al., 2013; Lockton et al., 2016), evolved to the 8 lenses of Architectural, Error-proofing, Interaction, Ludic, Perceptual, Cognitive, Machiavellian, and Security (Lockton, 2017). Table 5.1 presents the main themes of each of the lenses.

The DwI method follows the design pattern approach (Lockton et al., 2013). The design pattern approach has been influential within the Human-Computer Interaction community (Norman, 1999; Crumlish & Malone, 2009), and seeks to anchor thinking on a particular topic area or subject. The design pattern approach however has not primarily been used as tools for novel idea generation. Nevertheless, as (Lockton et al., 2010; Lockton et al., 2013) argue, such an approach can merit the consideration of multiple answers to set problem and, therefore, act to encourage multiple ideas. The design pattern approach as used within the DwI design cards comprises of a short title, labelling the card, colour coded to its particular lens, a question the designers can ask of their system, and a photograph of a prototypical example of the design card in action. Users are required to use the information presented on each of the cards to make their own inferences regarding their products and their end-users needs, with no boundaries set in place by the tool. Previous uses of the DwI approach include proposed redesign of automated teller machines (ATMs) (Lockton et al., 2010), with the authors suggesting that the use of DwI approach led to a variety of feasible design solutions. The DwI cards were also used by Salmon et al. (2018) as

TABLE 5.1
Summary of the Lenses and Themes Present within the DwI Framework

Lens	Theme	Example Cards	Number of Cards
Architectural Lens	Draws primarily on ideas within architecture and urban planning, seeking to apply ideas from the built environment. Largely concerns the structure and layout of items and behaviour.	Angles Pave the Cowpaths	12
Error-proofing Lens	Considers any behaviour that deviates from a target behaviour as an error and seeks to reduce the likelihood of errors occurring. Seeks to design a system whereby these errors cannot occur.	Are You Sure? Matched Affordances	10
Interaction Lens	Fundamentally about user's interaction with the devices or displays. Based on the feedback and feedforward of information between the user and the device being considered.	Kairos Real-Time Feedback	10
Ludic Lens	Focus on the potential for gamification of a device. Popularised by the view that playful interactions can encourage the maintenance of behaviour.	Scores Storytelling	11
Perceptual Lens	Seeks to utilise biases in human perceptual system, for example use of heuristics, to target the design and development of objects.	Colour Associations Nakedness	17
Cognitive Lens	Based on cognitive psychology and an understanding of how individuals make decisions. Seeks to bias individuals to make a desired decision.	Provoke Empathy Commitment and Consistency	15
Machiavellian Lens	Seeks to control the behaviour of individuals, by utilising an "Ends Justify the Means" approach.	Functional Obsolescence I Cut, You Choose	14
Security Lens	Seek to prevent undesired behaviour through direct countermeasures. Seeks to directly control behaviour.	Peervailence Coercive Atmospherics	12

part of a sociotechnical systems-based design process to create a series of new road intersection concepts. The aim was to develop new designs that would enhance interactions between drivers, cyclists, motorcyclists, and pedestrians in order to reduce collisions. Three novel intersection concepts were produced and evaluated with end-users from each road user group (Read et al., 2018; Salmon et al., 2018).

Rationale Summary

Previous arguments have stated that greenhouse gases and other pollutants released from everyday driving have significant negative environmental and health implications (Buckeridge et al., 2002; Thornton & Covington, 2015). As these emissions are a leading contributor to overall pollution (Hill et al., 2012), finding ways in which to minimise them are a key challenge for researchers across multiple domains and disciplines. One role that ergonomists and designers can take in this challenge is assisting in the development of in-vehicle interfaces designed to encourage fuel-efficient driving. One tool produced to assist designers and researchers in the development of such interfaces is the DwI toolkit (Lockton et al., 2010). The current study used a group workshop approach in order to explore the ideas that could emerge using the DwI approach when applied to developing interfaces encouraging fuel-efficient driving.

CASE STUDY

Participants

The workshop comprised 8 participants, 3 female participants aged 26–43 years ($M = 33.67$, $S.D. = 8.62$) and 5 male participants, aged 28–57 years ($M = 38.2$, $S.D. = 13.44$). Participants were from a variety of academic backgrounds including Psychology, Human Factors, Control Engineering, Automotive Engineering, and Electronic Engineering. Participants within this session all had an interest in improving vehicle fuel efficiency and were invited to participate in the workshop to take advantage of their specialist knowledge of vehicle systems, including knowledge of vehicular powertrains, which could inform potential interface development. This study was approved by the Ethics Committee, University of Southampton (ERGO 27172).

Procedure

This investigation utilised a case study approach, organised as a workshop event. At the start of the workshop, all participants were introduced to the goal of developing interfaces to support fuel-efficient driving. To assist this goal, participants were introduced to the concept of eco-driving and given the opportunity to discuss with the research team any questions they had regarding this concept, interactions between eco-driving and vehicles' mechanical systems, or the wider purpose of the workshop.

The workshop lasted approximately five hours. The first hour was devoted to the introduction and general discussion of the topic of fuel efficiency and eco-driving, and how each of the participants' diverse expertise could assist in this goal. During this introduction, participants were fully introduced to the structure of the workshop, and what was expected of them. In addition, it was clarified that an interface was not limited to any specific sensory modality, but was any system which promoted fuel-efficient driving. To assist with this, large posters were placed at the front of the room showing the phrases "Visual (Head Up Display, HUD)", "Visual (Head Down Display, HDD)", "Auditory", "Haptic (Pedals)", "Haptic (Seat)", and "Other". These posters would be used within the workshop to categorise participants' ideas for an interface into a specific feedback modality, but also acted to remind participants that additional implementation techniques were available beyond a visual display. Three hours, consisting the majority of the workshop was dedicated to the process of ideation, and discussion of the individual DwI cards. The final hour of the workshop was dedicated to a review of the generated ideas.

Following the introduction, each workshop participant was given a unique and random selection of 12 or 13 DwI cards, so that all 101 DwI cards were distributed between the workshop attendees. Each participant was then asked to sequentially lead a group discussion on one of the DwI cards they had been assigned, generating ideas relating to the development of new in-vehicle interfaces to support fuel-efficient driving. All participants were encouraged to contribute any ideas they had related to the presented card. In this way, every DwI card had the potential to inspire multiple design ideas in the form of a round-table discussion. Once an idea was generated, it was categorised as suitable for either a singular or multiple interface modality, and placed under the relevant marker at the front of the room, using a series of post-it notes. Participants were encouraged to be as creative as possible with their ideas throughout the workshop and were told that there were no bad ideas. Due to the nature of the DwI toolkit and the workshop's stance on encouraging the free creation of ideas, it was possible for the same, or highly similar, ideas to emerge from different DwI cards. If this occurred participants were encouraged to still record the idea. Once an idea had been suggested as part of the workshop, it was written on a post-it note and attached to its relevant feedback modality, visible to the workshop attendees. This process was repeated following a round table discussion for all 101 DwI cards creating a collection of 138 generated ideas.

Once all 101 DwI cards had been discussed by the research team, participants were invited to review the produced ideas, and encouraged to present any further ideas which they had relating to the interface development. Although no new ideas emerged as part of this review, several ideas created earlier within the workshop were clarified with clearer meaning to aid future transcription.

REVIEW OF THE IDEAS AND FINAL CODING

Following the workshop, the research team reviewed the ideas generated within the workshop, independent of the wider attendees. In order to make best use of the generated ideas within further research and in order to be suitable for the design of workable interfaces, the generated ideas were required to be reduced to a manageable number. Schunn et al. (2005) suggest that this process of refinement and reduction is a fundamental aspect of the design process. To achieve this goal, the number of ideas was culled in a two-stage process.

For the initial cull, ideas were classified as suitable or not suitable for implementation within an in-vehicle interface, the focus of the current research. This was essential in removing ideas linked to governmental policy shifts and providing greater financial incentives to eco-drivers. This initial cull removed approximate half of the ideas generated within the workshop, however, these ideas were stored for future consideration.

For the second cull, each of the remaining ideas were then classified as either constraint-based or demand-based, following the principles of ecological interface design (Vicente & Rasmussen, 1992). Constraint-based ideas sought to highlight a particular aspect of the fuel-efficient driving task to the driver in order to make the constraints that are present, but not obvious, more explicit. In contrast, demand-based ideas directly imposed a specific action onto the driver, demanding an appropriate action or forcing the car to react in a certain way. It was deemed prudent to pursue ideas that were classified as constraint-based for the design of vehicle interfaces in order to encourage potential user acceptance and the uptake of the developed ideas. Following this review, the remaining ideas were collected to produce initial designs.

DESIGN RESULTS & DISCUSSION

From the workshop, a total of 138 ideas were generated, highlighting that each of the 101 design cards within DwI has the potential to inspire multiple ideas. Importantly, all of the eight different lenses that make up the DwI toolkit contributed to the generated ideas, suggesting the lenses are not domain specific. Figure 5.1 presents the relative breakdown of the ideas across the different lenses of the framework, although the total number of available cards differs in each lens, it can be seen that all lenses contributed to the process of generating novel ideas.

Due to space limitations, not all 138 ideas generated within the workshop will be presented within the current chapter. Table 5.2 presents ideas that were generated using

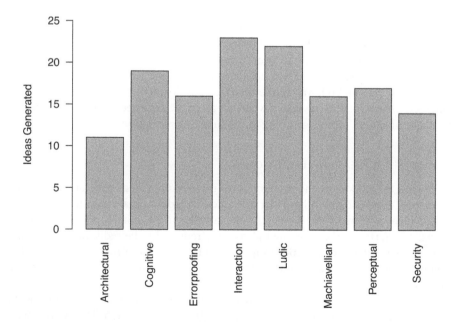

FIGURE 5.1 Number of ideas generated for each of the DwI Lens.

TABLE 5.2
Ideas Inspired by the Architectural Lens, including the Specific Cards Used

Feedback Category	Lens	Card	Description/Reasoning
Haptic-Pedal	Architectural	Roadblock	Limit the amount of acceleration possible via use of counterforce in pedal.
Haptic-Seat	Architectural	Material Properties	Artificial bumpiness, simulating harder suspension if drivers drive in a non-eco way.
Visual – HDD	Architectural	Angles	Make it easier to select eco-friendly options and harder to select non-eco modes.
Visual – HDD	Architectural	Positioning	Start stop system at traffic lights which is difficult to turn off.
Visual – HDD	Architectural	Simplicity	Visual display of the financial benefits of eco-driving.
Other	Architectural	Pave the Cowpaths	Codify eco-routes within the system.
Other	Architectural	Converging and Diverging	Encourage greater platooning of vehicles for greater fuel economy.
Other	Architectural	Converging and Diverging	Encourage car sharing between people with similar start/end journeys.
Other	Architectural	Hiding Things	Hide sports mode to the driver so that it is harder to access.
Other	Architectural	Hiding Things	Disable sports mode if fuel economy drops below a set of thresholds.
Other	Architectural	Roadblock	Non-eco buttons harder to reach than eco buttons.

cards from the Architectural lens. Despite this lens having a focus on the physical properties of objects and being largely inspired by ideas relating to the design of large-scale spaces, it was able to inspire 11 ideas relating to greater vehicle fuel efficiency.

When considering the feedback modalities of the generated ideas, that is, how each generated idea would interact with the driver of the vehicle, six key frames were considered "Visual (HUD)", "Visual (HDD)", "Auditory", "Haptic (Pedals)",

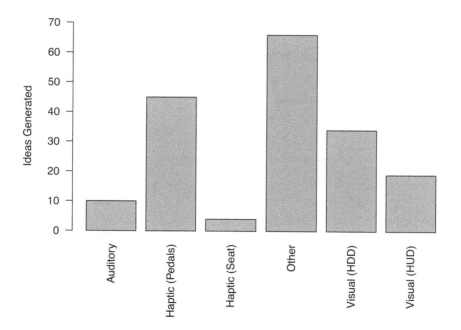

FIGURE 5.2 Number of ideas generated for each feedback modality during the workshop.

"Haptic (Seat)", and "Other". Figure 5.2 presents the distribution of ideas to each of the interface modalities that were considered as part of the workshop.

The research team did not anticipate the number of ideas that were classified as "Other". Generated ideas within this category included "Cheaper car based services for eco-drivers, for example variable parking rates based on fuel economy" and "Greater impact of driving style/eco-driving on insurance premiums". Although these ideas are valuable and have the potential for discussion and dissemination, they typically operate and a higher system level than the immediate driving environment, for example governmental policy and town planning, and are therefore not directly applicable to the challenges addressed within the current chapter, or indeed book, in designing interfaces to support fuel efficiency. Despite the potential value in these ideas, they were therefore removed within the initial cull. These ideas were, however, recorded for future analysis and interpretation.

As discussed within the "Methods" section of the current chapter, the research team was required to cull the ideas that had been generated in order to develop usable interfaces. For the first cull, all ideas that were not directly related to the development of potential in-vehicle interfaces were discarded. Primarily these ideas related to changing the physical characteristics of vehicles and promoting wider societal change and environmental awareness of the emissions released when driving. Despite the potential benefits of these ideas, the research team was unable to currently test ideas such as "Make the fuel tank harder to fill, in order put barriers to refueling". This initial cull resulted in 62 ideas being put aside. Due to the initial interface categorisation, the significant majority of these ideas were drawn from the "Other" category.

Despite this initial culling of ideas, a total of 76 ideas remained. Following the principles of ecological interface design as outlined previously, the research team completed the second cull. Following this process, it was found that 21 ideas were constraint-based and ecological, with the potential to be taken forward for the development of in-vehicle interfaces. Of the 21 ideas that were taken forward, 5 related to providing drivers with a visual gamification of fuel efficiency, such as a growing flower bed which flourishes or wilts based on driver performance or a polar bear sitting on a growing and shrinking iceberg which changes subject to the fuel efficiency of drivers' actions. Although both of these ideas are unique and potentially have different implications long-term, for the purpose of the current work, they were collated into a single idea, of a visual display encouraging gamification. Similarly, four generated ideas related to promoting knowledge of and encouraging the coasting behaviour, so they were also compiled into a single idea. Following this final review, 14 unique ideas were considered suitable for the development of user interfaces. The compiled interface ideas, as well as how they are intended to interact with and feedback information to the driver of the vehicle, are presented in Table 5.3.

DRIVER ACCEPTANCE

Following the development of the DwI informed interface designs, it was deemed prudent to explore the extent to which the different proposed interfaces would likely be both accepted by drivers and perceived as effective. This would act as an initial validation activity and offer researchers an insight into the interventions that would be of most value to consider within future empirical work. Usability has been considered key in previous work when considering the acceptance of in-vehicle interfaces (Harvey et al., 2011; Stanton & Salmon, 2011). Whilst a panel of eco-driving experts might be considered an ideal sample for this activity, it was reasoned by the research team that such experts would have limited need for the proposed interfaces generated with the current research. In contrast, drivers with no eco-driving training and no particular interest in fuel-efficient driving could significantly benefit from the potential savings induced by the system (McIlroy & Stanton, 2017a). Consequently, drivers with no previous eco-driving training were sought and recruited to assess the initial usability of the proposed interfaces.

VALIDATION METHODOLOGY

PARTICIPANTS

24 participants, (13 Male participants and 11 Female participants) completed the study validation questionnaires. Participants were recruited using opportunity sampling and were not compensated for their time spent completing this research. All participants were required to have held a current full driving license for a minimum of two years to be eligible to participate in the evaluation exercise. Participants who had extensive knowledge of or had received formal training in eco-driving practices were not eligible to take part.

TABLE 5.3
Summary of the 14 Interface Ideas Generated within the Workshop, Designed to Increase Drivers Fuel Efficiency

No.	Interface Idea	Auditory	Haptic Seat	Haptic Pedal	Visual (HUD)	Visual (HDD)
1.	Provision of a warning regarding the best moment to remove foot from accelerator to minimise required braking.					
2.	Tailored eco-driving suggestions based on location, e.g. notification regarding hidden bends and approaching traffic.					
3.	Colour changing lighting within the vehicle cabin based on drivers' fuel efficiency and recent actions.					
4.	Provision of an eco-driving and environmental start-up notification, promoting eco-driving, and highlighting the negative environmental effects of car use.					
5.	Visual display encouraging gamification of fuel economy. Polar bear on an iceberg shrinking and growing based on driver's fuel efficiency and recent actions.					
6.	Behavioural metaphor displays to mimic journey parameters. Challenge of keeping an item in the centre of a container based on lateral and longitudinal movements of the vehicle.					
7.	Impose a limit on the total possible acceleration should braking be shortly required.					
8.	Provision of a warning when a stop will be required, for example as a result of traffic light or congestion.					
9.	Dynamic fuel tank display that grows in prominence as fuel is depleted, highlighting the limited fuel reserves to the driver.					
10.	Dynamic band on the speedometer displaying best possible speed for the current location to minimise likelihood of stopping (traffic lights or congestion).					
11.	Dynamic range display and estimate based upon driver's fuel efficiency and recent actions.					
12.	Notification provided to the driver when excessive and unneeded acceleration is applied.					
13.	Fuel tank display based on money available rather than fuel available.					
14.	Display the financial cost of a journey at the end of each journey, including money saved or wasted.					

Applying Design with Intent to Support Creativity

MEASURES

Participants completed two rating questionnaires as part of the validation activity. Both questionnaires utilised a five-point likert scale. For the first questionnaire, participants were presented with 14 interface ideas and asked to rate the likelihood they would be willing to use each interface on a scale of Very Unlikely (1) to Very Likely (5). For the second questionnaire, participants were asked to consider, irrespective of their previous responses, the perceived impact that each interface would have on overall fuel use, using a scale of Very Low Impact (1) to Very High Impact (5).

PROCEDURE

Prior to taking part in the validation activity, participants were informed that they would be asked to judge their acceptance and initial impression of a series of interfaces designed to reduce the fuel usage associated with everyday driving. Following acceptance of this, participants were presented with the questionnaires. The researcher verbally explained each concept and sought confirmation that the participant understood the functionality of the different interfaces before participants completed the questionnaires. Where appropriate, participants were asked to complete this task independently, and not to discuss their ratings with other participants. The researcher was present at all times and participants were encouraged to seek clarification from the researcher if they were unclear of the meaning or functioning of any of the interface items. The questionnaires took approximately ten minutes to complete. Ethical approval for this research was given by the University of Southampton, Ethics Committee (ERGO 40131).

VALIDATION RESULTS & DISCUSSION

Table 5.4 presents participants median and standard deviation scores regarding likelihood of use for the different interfaces developed following the DwI workshop. It can be seen that participants rated the majority of items as likely to use, at a median of 4 or in the case of "Display cost of a journey at the end of each journey, including money saved or wasted", even higher. It was seen, however, that the fine detail task of "Challenge of keeping an item in the centre of a container" was rated poorly compared to the other interface displays, receiving a median rating of 2.

Initial data exploration was completed via the use of a Box and Whisker plot, presented in Figure 5.3 This figure supported the view that participants' ratings of likelihood to use the different interface displays differed. To explore these differences in more detail, a Friedman's test was calculated. This test revealed that there was a significant difference in participants' likelihood to use the different interface displays, χ^2 (13) = 43.82, p = .01. This analysis was expanded via the use of Wilcoxon signed-rank tests, with the Bonferroni correction, as post hoc tests, based upon a visual inspection of Figure 5.3. Five Post hoc tests were calculated, comparing lowest ranked interface display, "Challenge of keeping an item in the centre of a container" (*Mdn* = 2) to interface displays which appeared to be consistently more highly ranked. Results indicated that "Tailored eco-driving suggestions based on

TABLE 5.4
Participants Median and Standard Deviation for Likelihood of Use for the Different Interface Designs Developed following the DwI Workshop (where 1 = Very Unlikely and 5 = Very Likely)

	Median	Interquarle Range
Warning on the best moment to remove foot from accelerator to minimise required braking.	4	2.75
Tailored eco-driving suggestions based on location.	4	1
Mood/colour changing cabin based on current fuel efficiency and recent actions.	3	2.75
Eco-driving and environmental start-up notification.	4	1
Visual display encouraging gamification. Polar bear on an iceberg shrinking and growing based on driver's fuel efficiency and recent actions.	3	2
Challenge of keeping an item in the centre of a container.	2	2
Limit possible acceleration if braking will be shortly required.	4	1
Warning when a stop will be required.	4	0
Dynamic fuel tank display, highlighting limited fuel.	4	0
Dynamic band on the speedometer displaying best possible speed for the current location to minimise likelihood of stopping (traffic lights or congestion).	4	2
Dynamic range display and estimate based upon driver's fuel efficiency and recent actions.	4	1.5
Notification when excessive acceleration is applied.	4	2.75
Fuel tank display based on money available within the tank rather than fuel available.	3.5	3
Display cost of a journey at the end of each journey, including money saved or wasted.	4.5	1.75

location" ($Mdn = 4$, $Z = 3.75$, $p < .001$, $r = .77$), "Eco-driving and environmental start-up notification" ($Mdn = 4$, $Z = 2.89$, $p = .019$, $r = .59$), "Limit possible acceleration if braking will be shortly required" ($Mdn = 4$, $Z = 3.00$, $p = .013$, $r = .61$) "Dynamic range display and estimate based upon driver's fuel efficiency and recent actions" ($Mdn = 4$, $Z = 3.81$, $p < .001$, $r = .77$) and "Display cost of a journey at the end of each journey, including money saved or wasted" ($Mdn = 4.5$, $Z = 3.54$, $p < .001$, $r = .72$) were all rated significantly more likely to be used.

In addition to participants rating their likelihood to use the different interfaces designed within the workshop, participants were asked to rate how much of an impact

Applying Design with Intent to Support Creativity

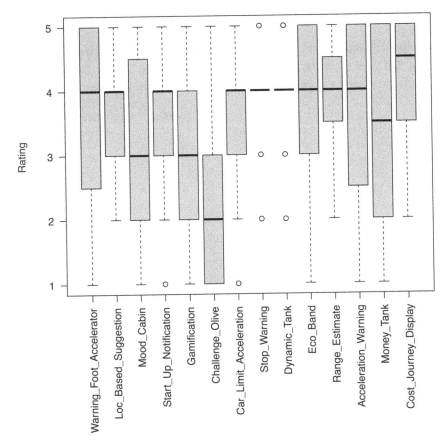

FIGURE 5.3 Box and Whisker plot for participants' likelihood to use the proposed interfaces.

the different interfaces would have on their overall fuel usage. Although no objective measures of fuel usage were taken within an empirical investigation and participants did not directly interact with the different interfaces, it was deemed important to consider the perceived benefits of each interface. Table 5.5 presents the median and standard deviations for these perceived effectiveness ratings. Whilst it was apparent that levels of perceived effectiveness were not as high as participant likelihood of use, numerous interface designs scored greater than 3, indicating that participants thought that such a display could make, regardless of size, a positive impact on overall fuel usage.

Similar to likelihood of use, the item rated as the least effective was the "Challenge of keeping an item in the centre of a container" which received a median rating of 2. Items regarded as more effective included "Warning on the best moment to remove foot from accelerator to minimise required braking", "Limit possible acceleration if breaking will be shortly required", "Dynamic band on the speedometer displaying best possible speed for the current location to minimise likelihood of stopping (traffic lights or congestion)", and "Display cost of a journey at the end of each journey, including money saved or wasted" all of which received a median rating of 4.

TABLE 5.5
Participants Median and Standard Deviation for Perceived Effectiveness for the Different Interface Designs Developed following the DwI Workshop (where 1 = Very Low Impact and 5 = Very High Impact).

	Median	Interquartile Range
Warning on the best moment to remove foot from accelerator to minimise required braking.	4	2
Tailored eco-driving suggestions based on location.	3	2
Mood/colour changing cabin based on current fuel efficiency and recent actions.	3	1.75
Eco-driving and environmental start-up notification.	3	1.75
Visual display encouraging gamification. Polar bear on an iceberg shrinking and growing based on driver's fuel efficiency and recent actions.	3	2
Challenge of keeping an item in the centre of a container.	2	2
Limit possible acceleration if braking will be shortly required.	4	1
Warning when a stop will be required.	3	1
Dynamic fuel tank display, highlighting limited fuel.	3.5	1.75
Dynamic band on the speedometer displaying best possible speed for the current location to minimise likelihood of stopping (traffic lights or congestion).	4	2.5
Dynamic range display and estimate based upon driver's fuel efficiency and recent actions.	3.5	2.5
Notification when excessive acceleration is applied.	3.5	1.75
Fuel tank display based on money available within the tank rather than fuel available.	3	2.5
Display cost of a journey at the end of each journey, including money saved or wasted.	4	2

To explore whether there were differences in participants perceived effectiveness of the different interface displays, a Friedman's test was calculated. This test revealed that there was a significant difference in participants perceived effectiveness of the different interface displays, $\chi^2 (13) = 54.30$, $p = .01$. Similar to likelihood ratings completed previously, a Box and Whisker plot was used as an initial data exploration technique, presented in Figure 5.4. Building on the main analysis, three post hoc tests were calculated comparing the lowest ranked interface item, again "Challenge of keeping an item in the centre of a container" (Mdn = 2) to interface items rated consistently higher. Results indicated that "Limit possible acceleration if breaking will be shortly required" ($Mdn = 4$, $Z = 3.58$, $p = .001$, $r = .73$), "Warning when a stop will be required" ($Mdn = 3$, $Z = 3.19$, $p = .004$, $r = .65$), and "Display cost of a journey at the end of each journey, including money saved or wasted" ($Mdn = 4$, $Z = 3.59$, $p < .001$, $r = .73$) were all perceived to have a significantly greater effect.

Applying Design with Intent to Support Creativity

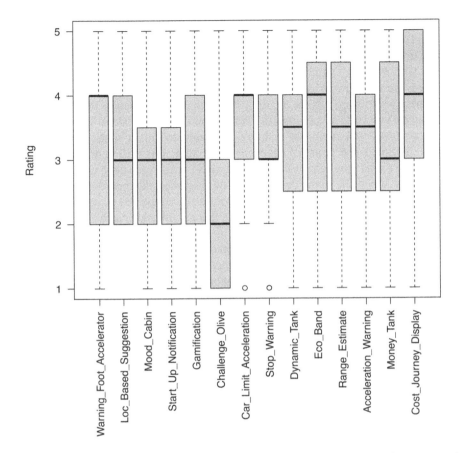

FIGURE 5.4 Box and Whisker plot for participants' perceived effectiveness of the proposed interfaces.

Overall results are supportive that participants would be willing to use the interfaces developed as part of the workshop. Despite general lower ratings for perceived effectiveness, combined these measures allow the focus of development for future laboratory studies of the proposed interventions. Upon reflection, qualitative data relating to why participants recorded the scores they did would have been beneficial for analysis. Not only would such data have yielded understanding regarding why participants left the ratings that they did, but such data would also have value in further developing and refining the interfaces prior to empirical testing using a driving simulator.

GENERAL DISCUSSION

The primary aim of this chapter was to use the DwI toolkit (Lockton et al., 2008; Lockton et al., 2010; Lockton et al., 2013; Lockton et al., 2016) to generate ideas to facilitate fuel-efficient driving, also referred to as eco-driving (Barkenbus, 2010).

Furthermore, this research sought to use the generated ideas to produce a collection of initial interface designs that could be matured into future full in-vehicle interfaces to support drivers and offer feedback in achieving greater fuel efficiency (Tulusan et al., 2012; McIlroy et al., 2013; Allison & Stanton, 2020). Using the DwI toolkit, 138 ideas were generated to encourage greater fuel efficiency within everyday driving. These ideas were generated using all of the available eight DwI lenses, and associated DwI cards to facilitate the generation of multiple ideas. Upon review of the ideas, it was apparent that many of the generated ideas suggested significant societal, political, and financial changes to encourage this shift in driver behaviour, which were outside the scope of interface design in this case study. Of the ideas that were highly applicable to the development of interfaces, the themes of gamification and increasing awareness of coasting behaviours were repeated across multiple suggestions. Previous research has demonstrated that gamification can be highly effective at reducing fuel usage, provided it is well designed and integrated with the overall driving experience. Dahlinger et al. (2018) explored the potential of numeric, compared to symbolic eco-driving notifications at reducing fuel usage based on an extensive sample of Swiss drivers, and found that symbolic notifications such as a growing tree displayed on the vehicle's dashboard reduced fuel usage by up to 3% as drivers attempted to minimise their fuel consumption. Similarly, promotion of coasting behaviours has been previously promoted within the literature to improve fuel efficiency. Beusen et al. (2009) explored the effect of an eco-driving training course and found greater coasting behaviours was associated with reduced fuel usage, and that coasting behaviours significantly increased following eco-driving training. Following consideration of the idea generated within the workshop, 14 concepts for potential interfaces were created.

Use of the DwI Toolkit

Of central interest to the current chapter is considering the value that design methods (Daalhuizen, 2014; Tromp & Hekkert, 2016), and specifically the DwI toolkit (Lockton et al., 2010), can play in assisting users to develop ideas and structure thinking on a particular topic. In the case of the current study, this was encouraging the adoption of eco-driving (Barkenbus, 2010), and similar fuel-efficient driving behaviours. Due to the scale of vehicle usage, and the negative environmental impact vehicle emissions can have when considering anthropogenic climate change and global weather patterns (Karl & Trenberth, 2003; Chapman, 2007), each small saving made by individuals can have considerable positive impact on the planet as a whole.

Whilst not a formal experiment, the current work has highlighted the potential of the DwI toolkit. The number of ideas generated within the workshop can demonstrate the immediate value of the DwI toolkit. The utility of the DwI approach is further supported by the relative distribution of ideas across lenses and that each DwI lens contributed new and innovative ideas. The value of the "suggestion tool" approach is further supported by the fact that the workshop attendees were not designers, but rather experts in Automotive Engineering and Human Factors. The DwI toolkit, therefore, presented an efficient approach to design, whereby individuals without extensive design experience were able to generate specifications, objectives, and initial interface designs. Previous research has demonstrated that design is a process that is fundamentally different from science and engineering, suggesting that access to the DwI toolkit facilitated creative thinking (Lawson, 1979). This dichotomy can be further magnified by the fact that participants were given complete freedom in the ideas that could be generated. Considering that the problem of addressing vehicle emissions and fuel usage can be seen as, to use design theory language, a "wicked problem" (Buchanan, 1992). Defined by Rittel and Webber (1973), wicked problems are *"a class of social system problems which are ill-formulated, where the information is confusing, where there are many clients and decision makers with conflicting values, and where the ramifications in the whole system are thoroughly confusing"*. Despite this difficulty, it was seen that the DwI toolkit offered a workable perspective to address the problem of improving fuel efficiency.

REALISING ECO-DRIVING THROUGH DESIGN

Brown (2009) proposed that *"Design thinking relies on our ability to be intuitive, to recognize patterns, to construct ideas that have emotional meaning as well as functionality, to express ourselves in media other than words or symbols"*. Taken within these confines, developing ideas and novel interface designs to facilitate fuel efficiency can be seen as design problems. Previous research has demonstrated the value that interfaces displays can have in reducing energy usage both within the home (Abrahamse et al., 2005; Jain et al., 2012) and within vehicles (Jamson et al., 2013; McIlroy et al., 2017). By considering ideas generated within the workshop from a constraint perspective as proposed by ecological interface design (Vicente & Rasmussen, 1992), the final interface designs generated from the workshop ideas were suitable for application within a vehicle.

Of key importance within design work targeting behavioural change is the effectiveness of the intervention. Tromp and Hekkert (2016) argue that within the field of design there is a frequent lack of agreement when judging the quality of a design outcome. Indeed, within the current work, the validation activity identified significant differences in participants' responses to the different interface design ideas. Extensive testing is required to address this current lack of clarity relating to both likelihood of use and perceived effectiveness. Once completed, this work will allow a bridge to be formed between the work of the designer and the work of the scientist and engineer (Lawson, 1979). This chapter has focused on the process of ideation and the initial development of interfaces to support fuel-efficient driving. The next steps of this research would be to empirically assess the different interfaces effectiveness at reducing fuel usage within a driving simulator. This research would help further validate the insights and value of the DwI approach.

CONCLUSION

This chapter has focussed on the role of a design method, specifically the DwI toolkit, in facilitating creativity in the generation of novel design ideas. Applied to the challenge of fuel-efficient driving, the DwI cards were first used to generate novel ideas designed to increase fuel efficiency, before the generated ideas were distilled to produce early designs for potential in-vehicle interfaces. Although further refinement, digitisation, and empirical testing is required of the proposed interfaces, it was clear that DwI offered a suitable approach to novel ideation to assist in the creative design process. Despite the fact that the DwI approach is reliant on the skills, expertise, and creativity of the individuals involved in the process, the lack of prior design expertise did not limit the generation of ideas. Based on the success of this approach, it appears that all lenses within the DwI toolkit can offer scientists, engineers, and designers useful insights in developing prototype interfaces.

REFERENCES

Abrahamse, W., Steg, L., Vlek, C., & Rothengatter, T. (2005). A review of intervention studies aimed at household energy conservation. *Journal of Environmental Psychology, 25*(3), 273–291.

Alexander, C. (1977). *A pattern language: Towns, buildings, construction.* Oxford University Press, Oxford, UK.

Allison, C. K., & Stanton, N. A. (2019). Eco-driving: The role of feedback in reducing emissions from everyday driving behaviours. *Theoretical Issues in Ergonomics Science.* DOI:10.1080/1463922X.2018.1484967.

Allison, C. K., & Stanton, N. A. (2020). Ideation using the "Design with intent" toolkit: A case study applying a design toolkit to support creativity in developing vehicle interfaces for fuel-efficient driving. *Applied ergonomics, 84,* 103026.

Attari, S. Z., DeKay, M. L., Davidson, C. I., & De Bruin, W. B. (2010). Public perceptions of energy consumption and savings. *Proceedings of the National Academy of Sciences, 107*(37), 16054–16059.

Barkenbus, J. N. (2010). Eco-driving: An overlooked climate change initiative. *Energy Policy, 38*(2), 762–769.

Barker, R. G. (1968). *Ecological psychology: Concepts and methods for studying the environment of human behavior.* Stanford University Press, Palo Alto, California.

Beusen, B., Broekx, S., Denys, T., Beckx, C., Degraeuwe, B., Gijsbers, M., & Panis, L. I. (2009). Using on-board logging devices to study the longer-term impact of an eco-driving course. *Transportation Research Part D: Transport and Environment, 14*(7), 514–520.

Bin, S., & Dowlatabadi, H. (2005). Consumer lifestyle approach to US energy use and the related CO_2 emissions. *Energy Policy, 33*(2), 197–208.

Birrell, S. A., Young, M. S., & Weldon, A. M. (2013). Vibrotactile pedals: Provision of haptic feedback to support economical driving. *Ergonomics, 56*(2), 282–292.

Brown, T. (2009). *Change by design: How design thinking transforms organizations and inspires innovation.* New York: Harper Collins.

Buchanan, R. (1985). Declaration by design: Rhetoric, argument, and demonstration in design practice. *Design Issues, 2,* 4–22.

Buchanan, R. (1992). Wicked problems in design thinking. *Design Issues, 8*(2), 5–21.

Buckeridge, D. L., Glazier, R., Harvey, B. J., Escobar, M., Amrhein, C., & Frank, J. (2002). Effect of motor vehicle emissions on respiratory health in an urban area. *Environmental Health Perspectives, 110*(3), 293.

Cash, P., Gram Hartlev, C., & Durazo, C. B. (2017). Behavioural design: A process for integrating behaviour change and design. *Design Studies, 48,* 96–128.

Cash, P., Holm-Hansen, C., Olsen, S. B., Christensen, M. L., & Trinh, Y. M. T. (2017). Uniting individual and collective concerns through design: Priming across the senses. *Design Studies, 49,* 32–65.

Chan, C. C. (2007). The state of the art of electric, hybrid, and fuel cell vehicles. *Proceedings of the IEEE, 95*(4), 704–718.

Chapman, L. (2007). Transport and climate change: A review. *Journal of Transport Geography*, *15*(5), 354–367.

Crumlish, C., & Malone, E. (2009). *Designing social interfaces: Principles, patterns, and practices for improving the user experience*. O'Reilly Media, Inc, Sebastopol, California.

Daalhuizen, J. (2014). *Method usage in design: How methods function as mental tools for designers*. Delft: Delft University of Technology.

Dahlinger, A., Tiefenbeck, V., Ryder, B., Gahr, B., Fleisch, E., & Wortmann, F. (2018). The impact of numerical vs. symbolic eco-driving feedback on fuel consumption – A randomized control field trial. *Transportation Research Part D: Transport and Environment*, *65*, 375–386.

De Bono, E. (2017). *Six thinking hats*. Penguin, London, UK.

Eckert, C., & Stacey, M. (2000). Sources of inspiration: A language of design. *Design Studies*, *21*(5), 523–538.

Franke, T., Arend, M. G., McIlroy, R. C., & Stanton, N. A. (2016). Eco-driving in hybrid electric vehicles – exploring challenges for user-energy interaction. *Applied Ergonomics*, *55*, 33–45.

Gardner, G. T., & Stern, P. C. (2008). The short list: The most effective actions US households can take to curb climate change. *Environment: Science and Policy for Sustainable Development*, *50*(5), 12–25.

Gibson, J. J. (1986). *The ecological approach to visual perception*. Hillsdale, NJ: Erlbaum.

Harvey, C., Stanton, N. A., Pickering, C. A., McDonald, M., & Zheng, P. (2011). A usability evaluation toolkit for in-vehicle information systems (IVISs). *Applied Ergonomics*, *42*(4), 563–574.

Hill, N., Brannigan, C., Smokers, R., Schroten, A., Van Essen, H., & Skinner, I. (2012). EU Transport GHG: Routes to 2050 ii. *final project report funded by the European Commission's Directorate-General Climate Action, Brussels*.

IJsselsteijn, W., De Kort, Y., Midden, C., Eggen, B., & Van Den Hoven, E. (2006, May). Persuasive technology for human well-being: Setting the scene. In: *International conference on persuasive technology* (pp. 1–5). Berlin, Heidelberg: Springer.

Jain, R. K., Taylor, J. E., & Peschiera, G. (2012). Assessing eco-feedback interface usage and design to drive energy efficiency in buildings. *Energy and Buildings*, *48*, 8–17.

Jamson, A. H., Hibberd, D. L., & Merat, N. (2013). The design of haptic gas pedal feedback to support eco-driving. In: *Proceedings of the Seventh International Driving Symposium on Human Factors in Driver Assessment, Training, and Vehicle Design* (pp. 264–270). University of Iowa.

Karl, T. R., & Trenberth, K. E. (2003). Modern global climate change. *Science*, *302*(5651), 1719–1723.

Lawson, B. R. (1979). Cognitive strategies in architectural design. *Ergonomics*, *22*(1), 59–68.

Lockton, D. (2017). Design with intent and the field of design for sustainable behaviour. In: *Living labs* (pp. 75–88). Springer International Publishing, Switzerland.

Lockton, D., Harrison, D., & Stanton, N. (2008). Design with intent: Persuasive technology in a wider context. *Persuasive*, *5033*, 274–278.

Lockton, D., Harrison, D., & Stanton, N. A. (2010). The design with intent method: A design tool for influencing user behaviour. *Applied Ergonomics*, *41*(3), 382–392.

Lockton, D., Harrison, D., & Stanton, N. A. (2013). Exploring design patterns for sustainable behaviour. *The Design Journal*, *16*(4), 431–459.

Lockton, D., Harrison, D., & Stanton, N. A. (2016). Design for sustainable behaviour: Investigating design methods for influencing user behaviour. *Annual Review of Policy Design*, *4*(1), 1–10.

Lorf, C., Martínez-Botas, R. F., Howey, D. A., Lytton, L., & Cussons, B. (2013). Comparative analysis of the energy consumption and CO_2 emissions of 40 electric, plug-in hybrid electric, hybrid electric and internal combustion engine vehicles. *Transportation Research Part D: Transport and Environment*, *23*, 12–19.

McGrenere, J., & Ho, W. (2000). Affordances: Clarifying and evolving a concept. In: *Proceedings of the graphics interface.* Toronto: Canadian Human-Computer Communications Society. 179–186.

McIlroy, R. C., & Stanton, N. A. (2017). What do people know about eco-driving? *Ergonomics, 60*(6), 754–769.

McIlroy, R. C., Stanton, N. A., & Harvey, C. (2013). Getting drivers to do the right thing: A review of the potential for safely reducing energy consumption through design. *IET Intelligent Transport Systems, 8*(4), 388–397.

McIlroy, R. C., Stanton, N. A., Godwin, L., & Wood, A. P. (2017). Encouraging eco-driving with visual, auditory and vibrotactile stimuli. *IEEE Transactions on Human Machine Systems, 47*(5), 661–672.

McIlroy, R. C., & Stanton, N. A. (2017). *Eco-driving: From strategies to interfaces.* CRC Press, Boca Raton.

Mensing, F., Bideaux, E., Trigui, R., Ribet, J., & Jeanneret, B. (2014). Eco-driving: An economic or ecologic driving style? *Transportation Research Part C: Emerging Technologies, 38,* 110–121.

Norman, D. A. (1999). Affordance, conventions, and design. *Interactions, 6*(3), 38–43.

Pannells, T. C., & Claxton, A. F. (2008). Happiness, creative ideation, and locus of control. *Creativity Research Journal, 20*(1), 67–71.

Philipsen, R., Brell, T., Funke, T., Brost, W., & Ziefle, M. (2018, July). With a little help from my government – A user perspective on state support for electric vehicles. In: *International conference on applied human factors and ergonomics* (pp. 386–397). Cham: Springer.

Ramanathan, V., & Feng, Y. (2009). Air pollution, greenhouse gases and climate change: Global and regional perspectives. *Atmospheric Environment, 43*(1), 37–50.

Read, G. J. M., Salmon, P. M., Goode, N., & Lenné, M. G. (2018). A sociotechnical design toolkit for bridging the gap between systems-based analyses and system design. *Human Factors and Ergonomics in the Manufacturing and Service Industries, 28*(6), 327–341.

Redström, J. (2006). Towards user design? On the shift from object to user as the subject of design. *Design Studies, 27*(2), 123–139.

Revell, K. M. A., & Stanton, N. A. (2016). Mind the gap – Deriving a compatible user mental model of the home heating system to encourage sustainable behaviour. *Applied Ergonomics, 57,* 48–61.

Revell, K. M. A., & Stanton, N. A. (2014). Case studies of mental models in home heat control: Searching for feedback, valve, timer and switch theories. *Applied Ergonomics, 45*(3), 363–378.

Revell, K. M. A., & Stanton, N. A. (2018). Mental model interface design: Putting users control of home heating. *Building Research & Information, 46*(3), 251–271.

Rittel, H. W., & Webber, M. M. (1973). Dilemmas in a general theory of planning. *Policy Sciences, 4*(2), 155–169.

Salmon, P. M., Read, G. J. M., Walker, G. H., Lenne, M. G., & Stanton, N. A. (2018). *Distributed situation awareness in road transport: Theory, measurement, and application to intersection design.* Boca Raton, FL: CRC Press.

Schunn, C. D., McGregor, M. U., & Saner, L. D. (2005). Expertise in ill-defined problem-solving domains as effective strategy use. *Memory & Cognition, 33*(8), 1377–1387.

Simon, H. A. (1969). *The sciences of the artificial.* Cambridge, MA: MIT Press.

Sivak, M., & Schoettle, B. (2012). Eco-driving: Strategic, tactical, and operational decisions of the driver that influence vehicle fuel economy. *Transport Policy, 22,* 96–99.

Stanton, N. A., & Salmon, P. M. (2011). Planes, trains and automobiles: Contemporary ergonomics research in transportation safety. *Applied Ergonomics, 42*(4), 529–532.

Thatcher, A., Garcia-Acosta, G., & Lange Morales, K. (2013). Design principles for green ergonomics. *Contemporary Ergonomics and Human Factors, 1,* 319–326.

Thornton, J., & Covington, H. (2015). Climate change before the court. *Nature Geoscience*, *9*(1), 3–5.
Tidwell, J. (2005). *Designing interfaces*. Sabastopol: O'Reilly.
Tromp, N., & Hekkert, P. (2016). Assessing methods for effect-driven design: Evaluation of a social design method. *Design Studies*, *43*, 24–47.
Tulusan, J., Staake, T., & Fleisch, E. (2012, September). Providing eco-driving feedback to corporate car drivers: What impact does a smartphone application have on their fuel efficiency? In: *Proceedings of the 2012 ACM Conference on Ubiquitous Computing* (*pp.* 212–215). ACM.
Vandenbergh, M. P., & Steinemann, A. C. (2007). The carbon-neutral individual. *NYUL Rev*, *82*, 1673.
Vicente, K. J., & Rasmussen, J. (1992). Ecological interface design: Theoretical foundations. *IEEE Transactions on Systems, Man, and Cybernetics*, *22*(4), 589–606.
Wever, R., Van Kuijk, J., & Boks, C. (2008). User-centred design for sustainable behaviour. *International Journal of Sustainable Engineering*, *1*(1), 9–20.
Young, M. S., Birrell, S. A., & Stanton, N. A. (2011). Safe driving in a green world: A review of driver performance benchmarks and technologies to support 'smart' driving. *Applied Ergonomics*, *42*(4), 533–539.

6 Incorporating Driver Preferences into Eco-Driving Optimal Controllers

INTRODUCTION

Drivers' acceleration and braking behaviour has a significant effect on vehicle fuel consumption, a fact that was first recognised over four decades ago (Evans, 1978). Reduction of fuel consumption, and therefore of carbon dioxide (CO_2) emissions due to transport, has become a topical issue in recent years due to concerns about climate change (Vandenbergh et al., 2007). Economical driving behaviour, or "eco-driving", has been suggested as a method which can reduce CO_2 emissions from road vehicles by 10% with current technology by encouraging drivers to accelerate gently, to anticipate signals and traffic flow to avoid stops, to maintain an even speed, and to avoid idling (Barkenbus, 2010). This view has been reinforced by recent results of the naturalistic driving study UDRIVE, which suggest that braking behaviour, gear shifting, and the velocity choice on the motorway have effects on fuel consumption of greater than 10% (Heijne et al., 2017).

Training programs have been designed to encourage eco-driving behaviors with positive results (Zarkadoula et al., 2007), but after these programs, many drivers revert back to their original driving styles over time (Beusen et al., 2009). A further complication for eco-driving is that avoiding stops at intersections requires prediction of traffic flow and signal changes, which is difficult for humans. Introducing coasting behaviour before corners and junctions is known to reduce fuel consumption, but coasting effectively requires prior knowledge of the timing of signal changes and accurate models of vehicle fuel consumption and dynamics (Rakha & Kamalanathsharma, 2011). A potential remedy to both of these problems is to provide active feedback to the driver on their behaviour, for instance by using auditory, visual, or haptic human-machine interfaces (HMIs) within a vehicle (Azzi et al., 2011; Birrell et al., 2013). There is considerable research interest in improving these interfaces using methods such as V2X communication (Li et al., 2015) or machine learning (Ma et al., 2018).

To put the problem of minimising fuel consumption in a mathematical framework, the eco-driving problem can be formulated as an optimal control problem (Wang et al., 2014; Sciarretta et al., 2015). This requires modelling of the vehicle powertrain, losses due to aerodynamic drag and rolling resistance, and modelling of gear-shifting, leading to a complex mixed-integer nonlinear programming problem (Saboohi & Farzaneh, 2009) or a dynamic programming problem (Mensing et al., 2011). Although effective, these methods require large computational effort to compute the optimal velocity profile. Model predictive control (MPC) has been suggested as a computationally tractable alternative to solve the problem during driving (Kamal et al., 2010). If available, the incorporation of fuel consumption maps and knowledge of road grade to such predictive optimisations can considerably improve performance (Jin et al., 2016). Recently, researchers have also considered the effect of road curvature on eco-driving (Ding & Jin, 2018), where constraints on lateral acceleration cause the vehicle to reduce speed in curves.

User acceptance is a well-known issue with existing automotive advanced driver assistance systems (ADAS), for example for collision-warning systems (Parasuraman et al., 1997) that rely on knowledge of typical vehicle-following behaviour. Based on the large variations between drivers observed in naturalistic studies, one suggestion to improve user acceptance of ADAS is to make the system adaptive, adjusting to the driver by estimating parameters representing their driving style in real-time (Fleming et al., 2019a). For eco-driving assistance systems, this possibility has already been explored for the specific scenario of approaching intersections (Xiang et al., 2015). Several simplified models of car-following behaviour have already been developed for use in traffic simulations, such as the intelligent driver model (IDM) which uses a differential equation for velocity to model the motion of a vehicle with acceleration limits and a preferred velocity and following distance (Treiber et al., 2000). For curve driving behaviour, models of tolerable lateral accelerations while cornering have appeared in the human factors literature, such as the model due to Reymond and co-workers appearing in Reymond et al. (2001). From the point of view of adapting system parameters, real-time adaptation of the IDM has been demonstrated (Monteil et al., 2015) and the authors have recently developed an automated fitting method for models of cornering speeds (Fleming et al., 2019b).

In this chapter, we present a general method of accounting for the effects of driver following behaviour and cornering speed choice in an eco-driving assistance system. By considering the modelling of longitudinal driver behaviour in an optimal control framework, it is possible to trade-off the objectives and preferences of the driver with the objective of energy efficiency. This provides a tuning parameter that can give behaviour intermediate between the two extremes of economical driving and natural driving and gives a systematic way to account for driving styles in an eco-driving assistance system.

To be able to represent a variety of drivers who have different preferences with regard to acceleration, braking, and following distances, our driver model contains a small number of parameters that may readily be estimated from naturalistic driving data. To simplify this process, we choose the set of model parameters to match those of the IDM, for which parameter fitting procedures are well known. We present a methodology based on an optimal control formulation of the IDM and subsequent quadratic approximation near equilibrium points that allows us to give physical meaning to the otherwise-arbitrary weighting parameters appearing in the cost function. This leads to a model that gives similar behaviour to the IDM in car-following scenarios, but that can additionally handle position-dependent constraints on velocity resulting from cornering and is suitable for incorporating into an eco-driving optimal control problem (Fleming et al., 2020). We validate this model by comparing its predictions with real-world driving data and existing models of cornering and car-following.

Because the new model of driver behaviour is based on optimal control in which the driver chooses the minimum possible value of a cost function, we may interpret the cost as a measure of a driver's "dissatisfaction" with a particular vehicle trajectory and control history. Under this interpretation, the model yields the acceleration profile that is most satisfying to the driver in terms of speed, acceleration, and vehicle spacing. Accordingly, we refer to it as the "Driver Satisfaction Model" (DSM) to easily contrast it with, for example, the IDM.

LITERATURE REVIEW

OPTIMAL CONTROL

Optimal control is an extension of the calculus of variations which considers how to choose the input of a dynamical system to optimise some performance criterion. The field originated with the work of Bellman (Bellman, 1957) and Pontryagin (Pontryagin et al., 1962) and was initially applied to problems in aeronautics and spaceflight, but later found application in many other fields. The optimal control problems we consider have the form:

$$\min_{\mathbf{u}(t) \in \mathcal{U}} J[\mathbf{x}, \mathbf{u}]$$

$$\text{s.t.} \quad \dot{\mathbf{x}} = f(\mathbf{x}, \mathbf{u}, t)$$

$$g(\mathbf{x}, \mathbf{u}, t) \leq 0$$

$$\mathbf{x}(0) = \mathbf{x}_0, \quad \mathbf{x}(T) \in \mathcal{X}_T$$

where $J[\mathbf{x}, \mathbf{u}]$ is the *cost function* given by:

$$J[\mathbf{x}, \mathbf{u}] = \int_0^T L(\mathbf{x}, \mathbf{u}, t) \, dt + \phi(\mathbf{x}(T)) \qquad (6.1)$$

In this cost function, $L(\mathbf{x}, \mathbf{u}, t)$ and $\phi(\mathbf{x}(T))$ are referred to as the *stage cost* and *terminal cost* respectively. The vector $\mathbf{x} \in \mathbb{R}^n$ represents the state of a dynamical system, which evolves according to the differential equation $\dot{\mathbf{x}} = f(\mathbf{x}, \mathbf{u}, t)$, and the vector $\mathbf{u} \in \mathbb{R}^m$ represents a control input. The cost function J models a quantity that should be kept small in a particular application, such as fuel usage, elapsed time, or deviation from a reference.

The problem of eco-driving can be formulated in an optimal control framework by choosing the stage cost $L(\mathbf{x}, \mathbf{u}, t)$ to be the rate of fuel consumption of the vehicle (Sciarretta et al., 2015). However, choosing to minimise fuel consumption alone can lead to behaviour that is unnatural for a human driver, such as travelling far below the speed limit or leaving large spacings to the preceding vehicle. To address this, we considered a modified optimal control problem in (Fleming et al., 2018) in which the cost function has the form:

$$J = \int_{T_i}^{T_f} \left(L_d + \alpha L_f \right) \mathrm{d}t \tag{6.2}$$

where L_f is a fuel consumption term, and the additional term L_d represents driver preferences on speed, acceleration, and inter-vehicle spacings. By adjusting the weighting parameter α between ∞ and 0, it is possible to obtain behaviour that is intermediate between minimising fuel consumption and fully respecting the driver's preferred driving style. This gives a parameter that can be tuned to ensure that the solution of the optimal control problem does not seem unnatural to the driver, respecting normal following distances for example.

Analytical solutions of optimal control problems only exist in special cases, such as when there are bounds on the input and both cost and dynamics are linear (Bellman et al., 1956) or for unconstrained linear systems with quadratic costs (Kalman, 1960). However, the latter case is practically useful because systems may often be linearised and cost functions locally approximated as quadratic, an approach that we adopt in the current work to relate the weighting parameters of our model's cost function to the parameters of existing models of driver behaviour. In practice, optimal control problems are often solved using numerical methods. Two effective approaches are multiple shooting, in which the input is parameterised as a function of time and the state integrated piecewise over times of interest (Bock & Plitt, 1984), and direct collocation, in which both the input and state are incorporated into a nonlinear programming problem as decision variables (Hargraves & Paris, 1987). Multiple shooting approaches are particularly useful when solutions to optimal control problems are required in real-time, such as for nonlinear model predictive control (Diehl et al., 2009). These methods of solution are not guaranteed to find the global solution of the problem, yet work well in practice.

Models of Vehicle Following

Many models of driver acceleration and braking behaviour express the acceleration of an individual driver as a function of their current speed and the relative position and velocity of the preceding vehicle. This leads to a differential equation describing motion known as a "car-following" model. Such car-following models have their origin in (Chandler et al., 1958), which proposes to describe vehicle-following by the delay differential equation:

$$\dot{v} = \lambda \left[v_L(t-\tau) - v(t-\tau) \right], \tag{6.3}$$

in which v denotes ego-vehicle velocity, v_L denotes the velocity of the leader, τ is a time-delay parameter, and λ is a sensitivity parameter. That is, the driver accelerates in proportion to the relative velocity of the vehicle ahead, incorporating some time delay. This shows a good fit to experimental data but is an oversimplified view of actual driver behaviour as there is no consideration of the distance to the lead vehicle. Subsequent works have incorporated this distance, as well as other factors, leading to many car-following models (Gipps, 1981; Bando et al., 1995; Krauß et al., 1997; Treiber et al., 2000). Because of the large number of models available in the literature, in the remainder of this section, we restrict our attention to two that are directly relevant to the present work. For a more extensive overview, we direct the reader to the review which may be found in Wilson and Ward (2011).

The IDM was proposed in Treiber et al. (2000) to model congested states of traffic, and is a car-following model described by the ordinary differential equation:

$$\dot{v} = a_{\max} \left[1 - \left(\frac{v}{v_d} \right)^\delta - \left(\frac{s^*(v, \Delta v)}{x_L - x} \right)^2 \right] \quad (6.4)$$

where the desired inter-vehicle spacing $s^*(v, \Delta v)$ is given by:

$$s^*(v, \Delta v) = s_0 + Tv + \frac{v \, \Delta v}{2\sqrt{ab}} \quad (6.5)$$

and where x_L denotes the position of the preceding vehicle and $\Delta v = v_L - v$ is the relative velocity of that vehicle. The model parameters $a_{\max}$, b, v_d, s_0, T may vary between drivers and are respectively the maximum acceleration, desired deceleration, free-flow velocity, minimum desired spacing, and the desired time-gap to the leading vehicle. Our interest in this model is that it provides a useful set of parameters that characterise driver behaviour during car-following, which have well-defined physical meanings and for which there exist well-tested fitting procedures (Kesting & Treiber, 2008).

A car-following model using linear quadratic optimal control, and hence that is related to the model developed in the present chapter, appeared in Burnham et al. (1974). This considered the driver as an optimal controller that regulates the vehicle spacing and velocity to desired set points s^* and v^* by minimising:

$$J = \int_0^\infty \left[q_s (s - s^*)^2 + q_v (v - v^*)^2 + \dot{v}^2 \right] dt \quad (6.6)$$

which, as there are no acceleration constraints, leads to linear driver behaviour according to the differential equation:

$$\dot{v} = -k_v (v - v^*) - k_s (s - s^*) \quad (6.7)$$

Although appealing due to its interpretation that the driver attempts to achieve preferred velocities and spacings to a lead vehicle, the model is unrealistic at large distances as then the acceleration is large and unbounded. A further practical issue is that the parameters q_s and q_v in the cost function have no physical interpretation, and instead the model must be fit to a particular driver by estimating k_s and k_v from driving data.

MODELS OF CORNERING SPEED

A model of driver speed choice in curves was suggested in (Godthelp, 1986) which, inspired by models of braking based on time-to-collision (TTC), considered a quantity called time-to-lane-crossing (TLC) to explain variations in driving speed for corners of different curvatures. By removing visual feedback while entering a corner, the authors demonstrated that drivers estimate the curvature of a corner on approach and make an anticipatory steering action based on the perceived curvature. Later work using a driving simulator also demonstrated that differences between individuals' cornering speeds could be explained by differences in their steering competence (Van Winsum & Godthelp, 1996), measured as their ability to accurately track the centre of a lane on a straight road.

Further study of driver speed in curves was carried out by Reymond et al. (2001), which investigated the effect of lateral acceleration in drivers' choice of speed when negotiating curves, demonstrating that drivers choose their cornering speed based on a lateral acceleration limit and a "safety margin" of error when estimating curvature. Denoting road curvature as κ, this leads to a model of maximum driver speed in the curve given by:

$$v_{max} = \sqrt{\frac{\Gamma_{max}}{\kappa + \Delta}} \tag{6.8}$$

where the parameters Γ_{max} and Δ represent the maximum lateral acceleration and the curvature safety margin. These parameters vary from driver to driver and together characterise a particular driver's cornering speed preferences. As it lacks a name, we refer to this model as "Reymond's model" for the remainder of the chapter.

A similar curvature-speed relationship is suggested in (Bosetti et al., 2015), where the maximum allowable speed in curves is predicted as:

$$v_{max} = \theta \kappa^{-\frac{1}{3}} \tag{6.9}$$

An interesting feature of that work is that a trajectory for vehicle longitudinal motion is generated by minimising a cost function that penalises time and includes a quadratic penalty on jerk, with this velocity limit included as a constraint. Hence, this models not only the choice of speed for a given curvature but also the transition while entering and exiting the corner. This is especially relevant to eco-driving applications, as coasting down to reduce speed before cornering can significantly affect fuel consumption.

THE DRIVER SATISFACTION MODEL

We now describe the driver satisfaction model (DSM) as a model of driver acceleration behaviour, noting that our goal is a model that gives similar behaviour to the IDM in car-following situations, but that can also handle velocity constraints due to cornering and be incorporated into eco-driving optimal control problems. In the following sections, we introduce a stage cost function for the DSM and then relate its weighting parameters to the parameters of the IDM. Variations in speed due to cornering are then incorporated by adding a constraint on vehicle velocity, in a similar manner to (Bosetti et al., 2015).

DEVELOPMENT OF COST FUNCTION AND CONSTRAINTS

The stage cost function used in the DSM is based on the optimal control model of car-following given in (Burnham et al., 1974), which has the cost function (6.6). Using this as a starting point, we make modifications to add a limit to the maximum acceleration and to modify the penalty on inter-vehicle spacing to better reflect the driver's indifference to far-away vehicles.

When the spacing s between vehicles is large, the behaviour of (6.6) becomes dominated by the $s - s^*$ term. This is quite unrealistic, as drivers are likely to be indifferent to far away vehicles. To address this, we introduce a penalty function for vehicle spacing s that tends to a constant value when s is large, reflecting this indifference. Hence, we use a penalty function that is approximately quadratic near its minimum but is a rational function with a quadratic term in the denominator according to:

$$\psi_s(s) = \frac{(s/s_d - 1)^2}{(s/s_d)^2 + 1} \tag{6.10}$$

in which s_d denotes the driver's "desired" spacing. The justification for referring to s_d as the driver's desired spacing is that it corresponds to the minimum of the penalty, and therefore will correspond to an equilibrium value for the model. The penalty is approximately constant for $s \gg s_d$, so that in this case the driver behaviour is unaffected by any lead vehicle.

To complete the cost, we add a quadratic penalty function for velocity that penalises the difference to a "desired" velocity which would be chosen by the driver in the absence of a lead vehicle. This is similar to (6.6). Combining these penalty functions then leads to the DSM stage cost:

$$L_{\text{dsm}} = (a/a_{\max})^2 + \beta(v/v_d - 1)^2 + \gamma \frac{(s/s_d - 1)^2}{(s/s_d)^2 + 1} \tag{6.11}$$

where β and γ represent weighting parameters to trade-off the three competing objectives.

Noting that minimisation of (11) alone may lead to unrealistically large accelerations when either $s - s^*$ or $v - v^*$ are large in magnitude, we also introduce a limit on the maximum acceleration. Considering the acceleration as a control input and denoting it by u, we limit it so that:

$$a \leq a_{\max} \tag{6.12}$$

where a denotes a constant parameter, which may be considered as a characteristic of a particular driver. We introduce a similar constraint on the vehicle spacing, specifying that:

$$s \geq 0 \tag{6.13}$$

at all times to avoid collisions with the lead vehicle.

CHOICE OF WEIGHTING PARAMETERS

At this point, the parameters s_d, β and γ are still arbitrary, making the model difficult to apply. Given the extensive literature on parameter estimation for car-following models, we relate these values to the parameters of the IDM. The following result provides a variational formulation of the IDM and will be used throughout the remainder of this section.

Theorem 1. *For fixed initial velocity $v(0) = v_i$, the IDM (6.4) is optimal for the stage cost*

$$L_{idm} = \left[\left(\frac{v}{v_d} \right)^\delta + \left(\frac{s^*}{s} \right)^2 - 1 \right]^2 + \left(\frac{a}{a_{\max}} \right)^2 - \left(\frac{a}{a_{\max}} \right) \left(\frac{s^*}{s} \right)^2 \tag{6.14}$$

and, writing $v_f = v(T)$, the terminal cost

$$\phi_{idm}(v_f) = \frac{2 v_f}{a_{\max}(\delta + 1)} \left[\left(\frac{v_f}{v_d} \right)^\delta - \delta - 1 \right] \tag{6.15}$$

subject to the constraints $\dot{v} = a$, $\dot{s} = v - v_L$.

Proof. From (6.4), the IDM is optimal for the stage cost

$$L_1 = \left[\frac{\dot{v}}{a_{\max}} + \left(\frac{v}{v_d} \right)^\delta + \left(\frac{s^*}{s} \right)^2 - 1 \right]^2$$

as this has a minimum value of zero which is attained if and only if (4) is satisfied at all times of interest.

Incorporating Driver Preferences into Eco-Driving Optimal Controllers

Expanding, collecting terms in $\dot{v}/a_{max}$, and integrating from $t=0$ to $t=T$, we find,

$$\int_0^T L_1 \, dt = \int_0^T L_{idm} \, dt + \int_0^T 2\left[\left(\frac{v}{v_o}\right)^\delta - 1\right]\frac{\dot{v}}{a_{max}} dt$$

Whereby the second term can be integrated by changing variables from t to v:

$$\int_{v_i}^{v_f} \frac{2}{a_{max}}\left[\left(\frac{v}{v_o}\right)^\delta - 1\right] dv = \phi_{idm}(v_f) - \phi_{idm}(v_i)$$

Because the initial velocity is fixed, the term $\phi_{idm}(v_i)$ is constant and may be dropped from the cost function without affecting the solution.

The cost function (14) is convenient for analysis, but not for direct minimisation via numerical methods. In particular, the first and second derivatives are equal to zero when $v = 0$ and approach zero as $s \to \infty$, which typically causes convergence problems. It also has a singularity at $s = 0$, which leads to ill-conditioning when s is small.

We construct expressions for s_d, β, and γ in terms of the parameters in the IDM by relating the two-stage costs (6.11) and (6.14) when considering small perturbations from equilibrium values. We first consider a vehicle on a straight road clear of other traffic. In that case, we may take $s \to \infty$ in the IDM stage cost (6.14):

$$L_{idm} \to \left[(v/v_d)^\delta - 1\right]^2 + (a/a_{max})^2$$

To fix β such that we obtain similar behaviour to (6.14) for v near v_d, we find a quadratic approximation $\tilde{q}(v)$ of:

$$q(v) = \left[(v/v_d)^\delta - 1\right]^2$$

by expansion in a Taylor series about v_d:

$$\tilde{q}(v) = q(v_d) + q'(v_d)(v - v_d) + \frac{1}{2}q''(v_d)(v - v_d)^2$$

Noting that $q(v_d) = q'(v_d) = 0$, we have:

$$\tilde{q}(v) = \frac{\delta^2}{v_d^2}(v - v_d)^2$$

after evaluating $q''(v_d)$. Comparing to the quadratic velocity penalty in the DSM stage cost (6.11), this suggests choosing:

$$\beta = \delta^2 \tag{6.16}$$

to obtain similar behaviour near equilibrium when s is large. With this choice of β we have:

$$\lim_{s \to \infty} \frac{\partial^2 L_{\text{dsm}}}{\partial v^2} \approx \lim_{s \to \infty} \frac{\partial^2 L_{\text{idm}}}{\partial v^2}$$

considering second derivatives at the equilibrium point $v = v_d$.

To find an expression for s_d, we consider the special case where there is a lead vehicle with a constant velocity v_d. Under these conditions, the IDM has an equilibrium at:

$$v_{\text{eq}} = v_L, \quad s_{\text{eq}} = \frac{s_0 + T v_L}{\sqrt{1 - (v_L/v_d)^\delta}} \tag{6.17}$$

By inspection of (6.11) we see that the minimum over s, and hence the equilibrium spacing when the lead vehicle is travelling at constant velocity, is at $s = s_d$. We would like this to correspond to the equilibrium of the IDM, so we choose:

$$s_d(v) = \frac{s_0 + T v}{\sqrt{1 - (v_L/v_d)^\delta}} \tag{6.18}$$

This gives greater following distances for larger values of v, in common with the IDM and many other models. The dependence on v rather than v_L is deliberate and leads to behaviour that is more similar to the IDM in braking, for which we recall $s^* = s_0 + Tv$.

To fix γ, we once again consider taking a local approximation of the IDM stage cost near this equilibrium point, requiring that the second partial derivatives with respect to s of L_{idm} and L_{dsm} are equal there. Hence, we require

$$\frac{\partial^2 L_{\text{dsm}}}{\partial s^2} \approx \frac{\partial^2 L_{\text{idm}}}{\partial s^2}$$

at the equilibrium point $v = v_{\text{eq}}$, $s = s_{\text{eq}}$. Evaluating the partial derivatives, we find:

$$\frac{\partial^2 L_{\text{dsm}}}{\partial s^2} = \frac{\gamma}{s_d^2} \tag{6.19}$$

Incorporating Driver Preferences into Eco-Driving Optimal Controllers

and:

$$\frac{\partial^2 L_{\text{idm}}}{\partial s^2} = \frac{8(s^*)^4}{s_d^6} + \frac{12(s^*)^2}{s_d^4} \left(\frac{(s^*)^2}{s_d^2} + \left(\frac{v}{v_0}\right)^\delta - 1 + \frac{a}{a_{\max}} \right) \quad (6.20)$$

To simplify (6.20), we note that if the DSM is to behave similarly to the IDM near equilibrium, then we will have:

$$\frac{a}{a_{\max}} \approx 1 - \left(\frac{v}{v_0}\right)^\delta - \frac{(s^*)^2}{s_d^2}$$

so that the second term may be neglected, leading to:

$$\frac{\partial^2 L_{\text{idm}}}{\partial s^2} = \frac{8(s^*)^4}{s_d^6} \quad (6.21)$$

Recalling that we wished the partial derivatives to be equal at the equilibrium point, we may now equate (6.19) and (6.21) and rearrange for γ to yield:

$$\gamma(v) = 8\left((v/v_d)^\delta - 1\right)^2 \quad (6.22)$$

This completes the specification of the stage cost function for the DSM, as the weighting parameters β, γ and the desired spacing s_d appearing in (6.11) have now been specified in terms of the driver preferences a, v_d, s_0, and T.

Remark 1. *If desired, the dependence of s_d and γ on v may be removed by substitution of $v = v_L$ which is true at equilibrium. In that case, the DSM stage cost (6.11) is quasiconvex in u, v, and s. This guarantees that it has a single local minimum which is also the global minimum. However, in cases considering limits on speed during cornering as in the following section, the resulting optimisation problem is nonconvex anyway and, in such cases, the authors have found that retaining the dependence on v typically leads to more natural driving behaviour.*

INCORPORATION OF CORNERING CONSTRAINTS

So far, the emphasis has been on developing a model compatible with the parameters of the IDM and yielding similar behaviour in vehicle-following. We now extend this to incorporate limits to the speed while cornering, considering the maximum speed as a function of curvature implied by the lateral acceleration limits of Reymond et al. (2001):

$$v \leq \sqrt{\frac{\Gamma_{\max}}{\kappa(x) + \Delta}} \quad (6.23)$$

We note that the curvature $\kappa(x)$ depends on the distance travelled down the road, and therefore the constraint function is, in general, a nonlinear function of travelled distance. Practically, this can be modelled as a cubic spline curve, which is the approach used in our implementation.

For completeness, we now state the entire model:

Optimisation 1 (Driver Satisfaction Model).
Solve the optimal control problem,

$$\min_{u(t)} \int_0^T L_{dsm}(x_L - x, v, a)\, dt$$

$$s.t. \quad \dot{x} = v, \quad \dot{v} = a$$

$$a \leq a_{max}, \quad x_L - x \geq 0,$$

$$v \leq \sqrt{\Gamma_{max}/(\kappa(x) + \Delta)}$$

$$x(0) = x_0, \quad v(0) = v_0$$

where the stage cost L_{dsm} is given by:

$$L_{dsm}(s, v, a) = (a/a_{max})^2 + \delta^2(v/v_d - 1)^2$$
$$+ 8\left((v/v_d)^\delta - 1\right)^2 \frac{(s/s_d - 1)^2}{(s/s_d)^2 + 1}$$

where we have defined:

$$s_d = \frac{s_0 + Tv}{\sqrt{1 - (v_L/v_d)^\delta}}$$

and where x_L is the position of the rear of the lead vehicle.

It should be noted that in this form, the model is acausal in that it assumes knowledge of the leader's future position $x_L(t)$ and the upcoming road curvature $\kappa(x)$. If it is desired to generate predictions from the model in real time, as is possible for the IDM, some simplifying assumption is required about the future behaviour of the lead vehicle, such as that it continues travelling at its current velocity, accelerates smoothly up to the speed limit, or that it brakes to a halt before an approaching intersection. When considering curve driving, however, acausal modelling is natural in that drivers decelerate in response to upcoming curves, before they are encountered.

Incorporating Driver Preferences into Eco-Driving Optimal Controllers

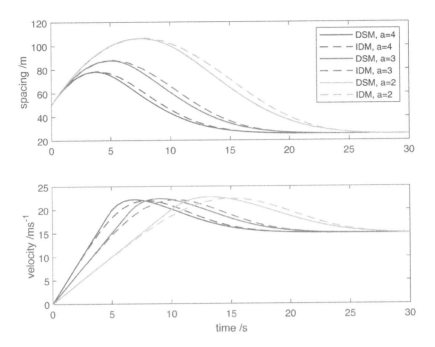

FIGURE 6.1 Car following, comparison with IDM for different 'a'.

COMPARISON OF THE MODEL WITH IDM

As the DSM expressions for s_d, β, and γ developed in the preceding sections were derived to provide similar behaviour to the IDM, it is interesting to compare the predictions of the two models in some common situations. Figure 6.1 shows a comparison of vehicle velocity under the DSM and the IDM when accelerating from a standing start to follow a vehicle of velocity 15 m/s, for different values of the parameter a. Other parameters used for this comparison are given in Table 6.1. The velocity and spacing of the two models evolve similarly in each case and increases in a (which represents the driver preference on maximum acceleration) affect both models by increasing the initial acceleration, decreasing the maximum spacing, and increasing the time taken to reach steady state, while leaving the maximum velocity almost unchanged.

TABLE 6.1
Model Parameters for the Comparison

Parameter	Value (IDM)	Value (DSM)
A	4 m/s²	4 m/s²
B	4 m/s²	–
v_d	30 m/s	30 m/s
s_0	2 m	2 m
T	1.5 s	1.5 s
δ	4	4

One interesting feature of the IDM compared to many other car-following models is the ability to handle emergency braking situations in which the deceleration may be large to avoid a collision. As the DSM contains no lower bound on the acceleration and contains a constraint to ensure that the vehicle spacing remains positive, it can also handle harsh braking situations in which large decelerations are required, avoiding a collision if possible to do so. A comparison between the two models under braking is shown in Figure 6.2, in which the ego-vehicle is initially travelling at 30 m/s and there is a slower vehicle travelling at 15 m/s at 50 m distance. In this figure, the parameter T is varied, which represents a driver preference on time headway while following another vehicle. In this case, the final spacing and velocity for the two models are equal, but the IDM shows much greater decelerations initially of over

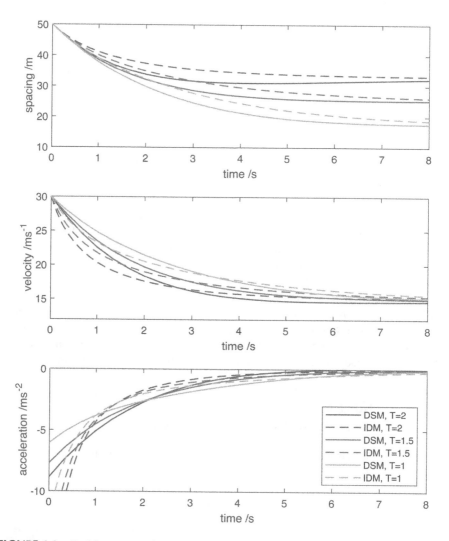

FIGURE 6.2 Braking, comparison with IDM for different 'T'.

10 m/s². For both models, the effect of increasing T is to increase the final following distance and to increase the initial deceleration, which is expected if a greater final following distance is desired.

MODEL VALIDATION

Method

We compared predictions of our model to naturalistic driving data collected by automotive data acquisition module (ADAM) (Yan et al., 2017) as part of the green adaptive control for interconnected vehicles (G-ACTIVE) project (Lot & Yan, 2016). ADAM is placed in a study participant's own vehicle and collects data on vehicle position and velocity using GPS as well as the spacing to the lead vehicle by post-processing of recorded stereo video. Data collected from three different drivers were used, which included both urban and rural driving situations. To give a fair evaluation of the two models, the data was split into "training" and "test" data. The training data consisted of approximately 10 minutes of car-following and cornering data for each driver that were used to fit parameters of the IDM and Reymond's cornering model by minimising the mean square velocity error, as described in Kesting & Treiber (2008) and Reymond et al. (2001). The parameters of the DSM were then chosen to be identical to the corresponding parameters of the IDM and the cornering model. The test data were separated into 30 second segments for use as cases for validation. Segments where the driver was travelling at a constant speed were discarded, and the remaining test cases were classified into: starting and stopping, vehicle-following, and cornering, by the following criteria:

- Start/stop: There is no leading vehicle. The driver either brakes to a stop at, or accelerates away from, a static road feature such as an intersection or traffic signals.
- Vehicle-following: The driver is following another vehicle, which is accelerating and braking due to traffic conditions. Corners and intersections may be included.
- Cornering: There is no leading vehicle. The driver's speed varies due to the road curvature at corners and intersections.

Examples of these three different cases are shown in Figures 6.3, 6.4, and 6.5 respectively. For the start/stop and vehicle-following tests, predictions of the IDM are compared to those of the DSM. Similarly, Reymond's model is compared with the predictions of the DSM for the cornering tests.

For each test case, both models were simulated using the leader velocity collected from ADAM and applicable speed limit and road curvature data. We computed the root-mean-square velocity error,

$$\text{RMSE} = \sqrt{\frac{1}{x(T)} \int_0^{x(T)} [v_{\text{predicted}}(x) - v_{\text{actual}}(x)]^2 \, dx}$$

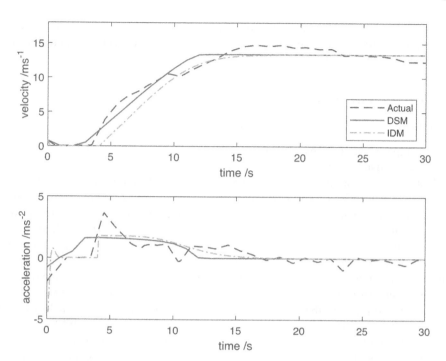

FIGURE 6.3 Example of start/stop test case (Test S4).

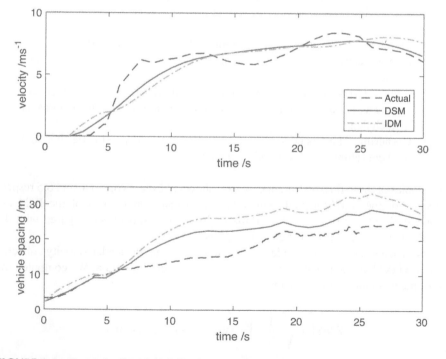

FIGURE 6.4 Example of vehicle-following test case (Test F4).

Incorporating Driver Preferences into Eco-Driving Optimal Controllers 133

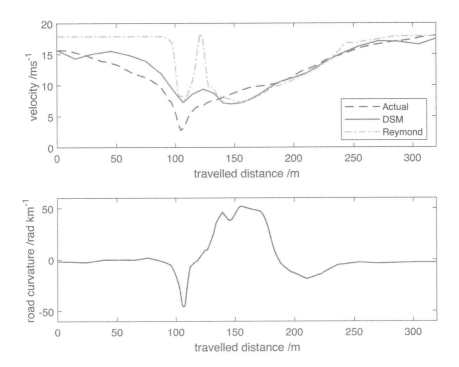

FIGURE 6.5 Example of cornering test case (Test C4).

and also the maximum velocity error,

$$\text{MAXE} = \max_{x \in [x(0), x(T)]} | v_{\text{predicted}}(x) - v_{\text{actual}}(x) |$$

to measure the goodness-of-fit of the model to the data. We take the integral with respect to distance in the root mean squared error (RMSE), rather than with respect to time, to ensure that we are always comparing speeds at the same position on the road. This is especially important for the cornering test cases.

Simulations and the subsequent calculation of error metrics were performed in MATLAB. The IDM was simulated by integrating the differential equations of motion using the Dormand-Prince scheme (MATLAB's *ode45* function) (Dormand & Prince, 1980). Reymond's model was evaluated by producing a static speed prediction as a function of travelled distance according to (6.8). The DSM was implemented by discretising the model for a step of 1*s* using direct collocation as described in (Hargraves & Paris, 1987). The automatic differentiation software CasADi (Andersson et al., 2012) was used to formulate the collocation equations, and the resulting nonlinear programming problem solved with Interior Point OPTimizer (IPOPT) (Wächter & Biegler, 2006). No attempt was made to improve computational efficiency (for example by code generation), yet this solution procedure took an average of 0.42s to optimise 30s of driver behaviour, using a mobile Intel core i5 processor, indicating that real-time implementation is likely to be feasible.

TABLE 6.2
Results for Start/Stop Test Cases

Test	$RMSE_{DSM}/ms^{-1}$	$RMSE_{IDM}/ms^{-1}$	$MAXE_{DSM}/ms^{-1}$	$MAXE_{IDM}/ms^{-1}$
S1	1.70	0.82	3.43	2.26
S2	0.57	0.84	1.17	2.07
S3	2.19	1.18	4.89	3.08
S4	0.97	1.22	1.77	3.44
S5	0.72	1.18	1.50	2.79
S6	0.73	1.55	1.91	4.28
S7	2.91	2.34	6.52	5.00
S8	3.03	2.31	7.10	4.99
S9	2.06	2.24	4.14	3.58
S10	1.06	2.73	2.34	4.31
Mean	1.59	1.64	3.48	3.58
Corr		0.48		0.60

RESULTS AND ANALYSIS

Result for all the testcases considered may be seen in Tables 6.2, 6.3 and 6.4 respectively. In all three scenarios, the RMSE and maximum error (MAXE) is smaller for the DSM than it is for either the IDM or Reymond's model, although in the car-following and start-stop test cases this improvement appears minor. All scenarios show a positive correlation between the error metric when simulating using the DSM and when using the reference model, implying that the DSM tends to perform well for the test cases in which the reference models perform well, and vice versa.

TABLE 6.3
Results for Vehicle-Following Test Cases

Test	$RMSE_{DSM}/ms^{-1}$	$RMSE_{IDM}/ms^{-1}$	$MAXE_{DSM}/ms^{-1}$	$MAXE_{IDM}/ms^{-1}$
F1	0.67	0.81	1.33	1.70
F2	1.31	0.65	1.87	3.70
F3	1.48	1.51	2.65	2.46
F4	0.82	1.11	2.08	2.70
F5	0.55	0.48	1.38	1.38
F6	0.30	0.32	0.71	0.87
F7	0.59	0.69	0.94	1.05
F8	0.34	0.47	0.81	1.01
F9	0.45	0.43	0.73	0.92
F10	1.48	2.80	3.11	4.34
Mean	0.80	0.93	1.56	2.01
Corr		0.78		0.89

Incorporating Driver Preferences into Eco-Driving Optimal Controllers

TABLE 6.4
Results for Cornering Test Cases

Test	RMSE$_{DSM}$/ms^{-1}	RMSE$_{Rey}$/ms^{-1}	MAXE$_{DSM}$/ms^{-1}	MAXE$_{Rey}$/ms^{-1}
C1	3.94	4.21	6.47	8.95
C2	2.74	4.93	4.11	9.17
C3	2.23	3.28	5.19	7.03
C4	1.84	3.68	5.99	10.34
C5	1.60	2.39	4.75	9.39
C6	2.90	3.61	8.83	12.85
C7	5.13	3.56	8.21	16.20
C8	1.07	1.51	2.83	4.10
C9	5.38	5.49	15.05	15.05
C10	3.08	6.35	4.65	12.35
Mean	2.99	3.90	6.61	10.54
Corr		0.58		0.73

To determine if these reductions in error were statistically significant, we compared the RMSE values for the DSM with that of the benchmark using the Wilcoxon signed-rank test, a non-parametric test of the paired differences. This indicated that the differences in RMSE for the start/stop ($Z = 0$, $p = 1$, $r = 0$) and vehicle-following ($Z = -1.27$, $p = 0.202$, $r = 0.40$) test cases were not significant at the 0.05 significance level. For the cornering test case, the differences in RMSE were significant ($Z = -2.08$, $p = 0.037$, $r = 0.66$), which was also the case for the MAXE error metric ($Z = -2.61$, $p = 0.009$, $r = 0.82$). This suggests that the DSM gives comparable performance to the IDM in car-following and start-stop situations while outperforming the model of Reymond et al. in cornering situations.

USAGE EXAMPLE

We now demonstrate the use of the DSM in an eco-driving optimisation by considering a braking scenario in which an electric vehicle must slow from highway speeds to a low speed due to a corner, as may occur at the end of an off-ramp or slip road. We consider a vehicle that may regenerate a factor $0 \leq \theta \leq 1$ of braking energy and has energy losses due to aerodynamic drag and rolling resistance in which the combined drag force is given by:

$$F_{drag} = c_d v^2 + mc_{rr} = \frac{1}{2}\rho_{air} C_D A v^2 + C_{RR} mg$$

in which $C_D A$ and C_{RR} denotes the usual coefficients of aerodynamic drag and rolling resistance, and mg is the weight of the vehicle. We additionally consider a simplified model of an electric powertrain with losses due to the resistance R_m of the motor windings and a reduction gear of ratio N_g. These resistive losses are given by $R_m i_m^2$ where i_m is the motor current given by:

$$i_m = \frac{rF}{N_g k}$$

in which F denotes the traction force at the wheels, r denotes wheel radius, and k is the torque constant of the electrical motor. For simplicity, we ignore other electrical losses.

To correctly handle the energy loss when braking, we introduce different input variables for acceleration due to the vehicle engine and brakes, considering $a = u_e + u_b$ where $a_e \geq 0$ is the component of acceleration due to the engine only, and $a_b \leq 0$ is that due to braking. Considering the a term in (6.11), we can expand in a_e and a_b giving:

$$(a/a_{\max})^2 = (a_e/a_{\max})^2 + 2a_e a_b/a_{\max}^2 + (a_b/a_{\max})^2$$

and noting that $a_e a_b = 0$ if, as in normal driving, the accelerator and brake are not used simultaneously, we find:

$$(a/a_{\max})^2 = (a_e/a_{\max})^2 + (a_b/a_{\max})^2 \tag{6.24}$$

In fact, as we may always establish this equality for any solution $u(t)$ of Optimisation 1 by choosing $a_e(t) = \max(0, a(t))$ and $a_b(t) = \min(0, a(t))$, it is evident that using the expression on the right-hand side of (6.24) in place of $(a/a_{\max})^2$ in the DSM stage cost will ensure that accelerator and brake are not activated simultaneously, without otherwise changing the result. This leads to the modified stage cost function:

$$\begin{aligned}L'_{dsm} &= (a_e/a_{\max})^2 + (a_b/a_{\max})^2 + \delta^2(v/v_d - 1)^2 \\ &+ 8\left((v/v_d)^\delta - 1\right)^2 \frac{(s/s_d - 1)^2}{(s/s_d)^2 + 1}\end{aligned} \tag{6.25}$$

For completeness, we state the optimal control problem with this modified stage cost.

Optimisation 2 (Example eco-driving optimisation).
Solve the optimal control problem,

$$\min_{u(t)} \int_0^T \left[L'_{dsm}(x_L - x, v, a_e, a_b) + \alpha L_{loss}(v) \right] dt$$

$$\text{s.t.} \quad \dot{x} = v, \quad \dot{v} = a_e + a_b - c_d v^2 - c_{rr}$$

$$0 \leq a_e \leq a_{\max}, \quad a_b \leq 0, \quad s, v \geq 0$$

$$v \leq \sqrt{\Gamma_{\max}/(\kappa(x) + \Delta)}$$

$$x(0) = x_0, \quad v(0) = v_0$$

where the DSM stage cost L'_{dsm} is given by

$$L'_{dsm} = (a_e/a_{max})^2 + (a_b/a_{max})^2 + \delta^2(v/v_d - 1)^2$$
$$+ 8\left((v_L/v_d)^\delta - 1\right)^2 \frac{(s/s_d - 1)^2}{(s/s_d)^2 + 1}$$

with $s_d(v)$ is defined as in Optimisation 1, and

$$L_{loss}(v) = -(1-\theta)ma_b v + c_d v^3 + c_{rr} v + R_m \left(\frac{rmu_e}{N_g k}\right)^2$$

is the stage cost associated with energy losses.

The result of this optimisation for different values of α in the braking scenario is shown in Figure 6.6. The corner is modelled as a cubic spline of increasing curvature as a function of x, with curvature increasing over an approximately 80 m length, and a final (maximum) curvature chosen to correspond to a cornering speed of 6 m/s. The initial velocity was taken as $v_0 = 25$ m/s, and the vehicle is initially at $x_0 = 0$ km with the corner placed at $x_c = 1.2$ km. For the vehicle, we take $m = 1500$ kg, $r = 0.29$, $C_D A = 0.7$, and $C_{RR} = 0.005$, typical for a medium-sized family car. For the electric powertrain we set $k = 0.12$, $R_m = 0.1$, $N_g = 15$, and $\theta = 0.7$. The driver preference parameters used in the DSM are those in Table 6.1 that were used for the previous comparison examples.

It is apparent that as the parameter α penalising energy usage is increased, the electric motor is used less and at lower torques on approach to the corner, with a greater reduction in speed via coasting under the effects of drag. In the most extreme case of $\alpha = 0.3$, this requires 900 m of coasting and reduces energy loss by approximately 43%, but intermediate values of α also give reductions in energy loss while remaining closer to the "natural" velocity profile given by the $\alpha = 0$ case.

CONCLUSION

We have introduced a method based on optimal control to incorporate driver preferences into eco-driving assistance systems. Models of longitudinal driver behaviour available from the traffic modelling literature were unsuitable for this purpose, so we developed a new model, the DSM, that characterises the driver using parameters in common with the well-known IDM, but in addition is compatible with eco-driving optimal control problems and is capable of reproducing driver speed in curves. As the DSM is based on optimal control, it may be directly used for eco-driving optimisations in which fuel-saving velocity profiles are sought while respecting a "natural" driving style. The model has been validated by comparison with naturalistic driving data in several scenarios, finding it gives comparable or better performance than existing models in the literature for prediction of the speed of an individual

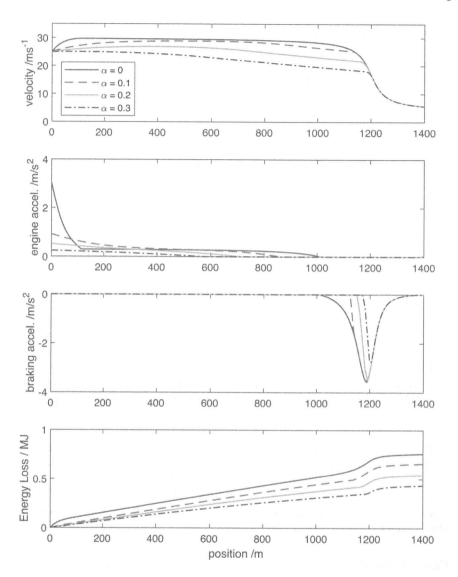

FIGURE 6.6 Eco-driving optimal control example.

vehicle in typical vehicle-following and cornering situations. Finally, we illustrated the potential for optimal control to be used to trade-off energy savings and a natural driving style in a braking and cornering scenario for an electric vehicle.

REFERENCES

Andersson, J., Åkesson, J., & Diehl, M. (2012). CasADi: A symbolic package for automatic differentiation and optimal control. In *Recent advances in algorithmic differentiation* (pp. 297–307). Springer, Berlin, Heidelberg.

Azzi, S., Reymond, G., Mérienne, F., & Kemeny, A. (2011). Eco-driving performance assessment with in-car visual and haptic feedback assistance. *Journal of Computing and Information Science in Engineering, 11*(4), 041005.

Bando, M., Hasebe, K., Nakayama, A., Shibata, A., & Sugiyama, Y. (1995). Dynamical model of traffic congestion and numerical simulation. *Physical Review E, 51*(2), 1035.

Barkenbus, J. N. (2010). Eco-driving: An overlooked climate change initiative. *Energy Policy, 38*(2), 762–769.

Bellman, R., Glicksberg, I., & Gross, O. (1956). On the bang-bang control problem. *Quarterly of Applied Mathematics, 14*(1), 11–18.

Bellman, R. E. (1957). *Dynamic programming.* Courier Dover Publications

Beusen, B., Broekx, S., Denys, T., Beckx, C., Degraeuwe, B., Gijsbers, M., Scheepers, K., Govaerts, L., Torfs, R. & Panis, L. I. (2009). Using on-board logging devices to study the longer-term impact of an eco-driving course. *Transportation research part D: transport and environment, 14*(7), 514–520.

Birrell, S. A., Young, M. S., & Weldon, A. M. (2013). Vibrotactile pedals: Provision of haptic feedback to support economical driving. *Ergonomics, 56*(2), 282–292.

Bock, H. G., & Plitt, K. J. (1984). A multiple shooting algorithm for direct solution of optimal control problems. *IFAC Proceedings Volumes, 17*(2), 1603–1608.

Bosetti, P., Da Lio, M., & Saroldi, A. (2015). On curve negotiation: From driver support to automation. *IEEE Transactions on Intelligent Transportation Systems, 16*(4), 2082–2093.

Burnham, G., Seo, J., & Bekey, G. (1974). Identification of human driver models in car following. *IEEE Transactions on Automatic Control, 19*(6), 911–915.

Chandler, R. E., Herman, R., & Montroll, E. W. (1958). Traffic dynamics: Studies in car following. *Operations Research, 6*(2), 165–184.

Diehl, M., Ferreau, H. J., & Haverbeke, N. (2009). Efficient numerical methods for nonlinear MPC and moving horizon estimation. In *Nonlinear model predictive control* (pp. 391–417). Springer, Berlin, Heidelberg.

Ding, F., & Jin, H. (2018). On the optimal speed profile for eco-driving on curved roads. *IEEE Transactions on Intelligent Transportation Systems, 19*(12), 4000–4010.

Dormand, J. R., & Prince, P. J. (1980). A family of embedded runge-kutta formulae. *Journal of Computational and Applied Mathematics, 6*(1), 19–26.

Evans, L. (1978). Driver behavior effects on fuel consumption in urban driving. In: *Proceedings of the Human Factors Society Annual Meeting*, volume 22, pp. 437–442. Sage Publications Sage CA: Los Angeles, CA.

Fleming, J., Yan, X., Allison, C., Stanton, N., & Lot, R. (2018, October). Driver modeling and implementation of a fuel-saving ADAS. In *2018 IEEE International Conference on Systems, Man, and Cybernetics (SMC)* (pp. 1233–1238). IEEE.

Fleming, J., Yan, X., & Lot, R. (2020). Incorporating driver preferences into eco-driving assistance systems using optimal control. *IEEE Transactions on Intelligent Transportation Systems, 22*(5), 2913–2922.

Fleming, J. M., Allison, C. K., Yan, X., Lot, R., & Stanton, N. A. (2019a). Adaptive driver modelling in ADAS to improve user acceptance: A study using naturalistic data. *Safety Science, 119*, 76–83.

Fleming, J. M., Yan, X., & Lot, R. (2019b) Fitting cornering speed models with one-class support vector machines. In: *2019 IEEE Intelligent Vehicles Symposium (IV)*, pp. 2457–2462. IEEE.

Gipps, P. G. (1981). A behavioural car-following model for computer simulation. *Transportation Research Part B: Methodological, 15*(2), 105–111.

Godthelp, H. (1986). Vehicle control during curve driving. *Human Factors, 28*(2), 211–221.

Hargraves, C. R., & Paris, S. W. (1987). Direct trajectory optimization using nonlinear programming and collocation. *Journal of Guidance, Control, and Dynamics, 10*(4), 338–342.

Heijne, V., Ligterink, N., & Stelwagen, U. (2017). Potential of eco-driving. UDRIVE Deliverable D45.1. EU FP7 Project UDRIVE Consortium. https://doi.org/10.26323/UDRIVE_D45.1.

Jin, Q., Wu, G., Boriboonsomsin, K., & Barth, M. J. (2016). Power-based optimal longitudinal control for a connected eco-driving system. *IEEE Transactions on Intelligent Transportation Systems, 17*(10), 2900–2910.

Kalman, R. E. (1960). Contributions to the theory of optimal control. *Bol. Soc. Mat. Mexicana, 5*(2), 102–119.

Kamal, M. A. S., Mukai, M., Murata, J., & Kawabe, T. (2010, September). On board eco-driving system for varying road-traffic environments using model predictive control. In *2010 IEEE International Conference on Control Applications* (pp. 1636–1641). IEEE.

Kesting, A., & Treiber, M. (2008). Calibrating car-following models by using trajectory data: Methodological study. *Transportation Research Record, 2088*(1), 148–156.

Krauß, S., Wagner, P., & Gawron, C. (1997). Metastable states in a microscopic model of traffic flow. *Physical Review E, 55*(5), 5597.

Li, S. E., Xu, S., Huang, X., Cheng, B., & Peng, H. (2015). Eco-departure of connected vehicles with v2x communication at signalized intersections. *IEEE Transactions on Vehicular Technology, 64*(12), 5439–5449.

Lot, R., & Yan, X. (2016). The G-ACTIVE project. http://www.g-active.uk, 2016. Accessed: 2018-04-13.

Ma, H., Xie, H., & Brown, D. (2018). Eco-driving assistance system for a manual transmission bus based on machine learning. *IEEE Transactions on Intelligent Transportation Systems, 19*(2), 572–581.

Mensing, F., Trigui, R., & Bideaux, E. (2011). Vehicle trajectory optimization for application in eco-driving. In: *Vehicle Power and Propulsion Conference (VPPC), 2011 IEEE*, pp. 1–6. IEEE.

Monteil, J., OHara, N., Cahill, V., & Bouroche, M. (2015). Real-time estimation of drivers' behaviour. In: *Intelligent Transportation Systems (ITSC), 2015 IEEE 18th International Conference on*, pp. 2046–2052. IEEE.

Parasuraman, R., Hancock, P. A., & Olofinboba, O. (1997). Alarm effectiveness in driver-centred collision-warning systems. *Ergonomics, 40*(3), 390–399.

Pontryagin, L. S. (1962). *Mathematical theory of optimal processes* (K. N. Trirogoff, Trans.). Interscience.

Rakha, H., & Kamalanathsharma, R. K. (2011). Eco-driving at signalized intersections using V2I communication. In: *Intelligent Transportation Systems (ITSC), 2011 14th International IEEE Conference on*, pp. 341–346. IEEE.

Reymond, G., Kemeny, A., Droulez, J., & Berthoz, A. (2001). Role of lateral acceleration in curve driving: Driver model and experiments on a real vehicle and a driving simulator. *Human Factors, 43*(3), 483–495.

Saboohi, Y., & Farzaneh, H. (2009). Model for developing an eco-driving strategy of a passenger vehicle based on the least fuel consumption. *Applied Energy, 86*(10), 1925–1932.

Sciarretta, A., De Nunzio, G., & Ojeda, L. L. (2015). Optimal eco-driving control: Energy-efficient driving of road vehicles as an optimal control problem. *IEEE Control Systems, 35*(5), 71–90.

Treiber, M., Hennecke, A., & Helbing, D. (2000). Congested traffic states in empirical observations and microscopic simulations. *Physical Review E, 62*(2), 1805.

Van Winsum, W., & Godthelp, H. (1996). Speed choice and steering behavior in curve driving. *Human Factors, 38*(3), 434–441.

Vandenbergh, M. P., Barkenbus, J., & Gilligan, J. (2007). Individual carbon emissions: The low-hanging fruit. *UCLA L. Rev, 55*, 1701.

Wächter, A., & Biegler, L. T. (2006). On the implementation of an interior-point filter line-search algorithm for large-scale nonlinear programming. *Mathematical Programming, 106*(1), 25–57.

Wang, M., Daamen, W., Hoogendoorn, S. P., & Van Arem, B. (2014). Rolling horizon control framework for driver assistance systems. Part I: Mathematical formulation and non-cooperative systems. *Transportation Research Part C: Emerging Technologies, 40*, 271–289.

Wilson, R. E., & Ward, J. A. (2011). Car-following models: Fifty years of linear stability analysis–A mathematical perspective. *Transportation Planning and Technology, 34*(1), 3–18.

Xiang, X., Zhou, K., Zhang, W. B., Qin, W., & Mao, Q. (2015). A closed-loop speed advisory model with driver's behavior adaptability for eco-driving. *IEEE Transactions on Intelligent Transportation Systems, 16*(6), 3313–3324.

Yan, X., Fleming, J., Allison, C., & Lot, R. (2017). Portable Automobile Data Acquisition Module (ADAM) for naturalistic driving study. In: *15th European Automotive Congress (EAEC 2017)*.

Zarkadoula, M., Zoidis, G., & Tritopoulou, E. (2007). Training urban bus drivers to promote smart driving: A note on a Greek eco-driving pilot program. *Transportation Research Part D: Transport and Environment, 12*(6), 449–451.

7 Receding Horizon Eco-Driving Assistance Systems for Electric Vehicles

INTRODUCTION

Environmental concerns and the drive towards greater energy efficiency poses great challenges for the automotive industry, making it more urgent to reduce the energy consumption of motor vehicles (Achour et al., 2011). There are two popular research trends with the aim to reduce the energy usage and cut emissions of road vehicles: a) the electrification of vehicle powertrains (Chen et al., 2015), and b) guiding the human drivers towards more environmentally friendly driving style referred to as eco-driving (Af Wåhlberg, 2006; Af Wåhlberg, 2007).

Electrified vehicles, such as battery electric vehicles (EVs) and hybrid electric vehicles (HEVs), are currently intensively investigated and manufactured by the automotive industry, and also receive profound focus in academic research (Ke et al., 2017). Recently, the marketability of electrified vehicles has become a widely discussed issue. Range anxiety and high cost make electrified vehicles less favourable for private users. Therefore, there is an ongoing effort to push for even higher energy efficient electrified powertrains and to make them more competitive. The energy efficiency of these vehicles could be improved by refining the motor design and choice of energy storage systems (Castaings et al., 2016). However, an essential and cost-effective measure is to improve the energy management systems, such that the powertrain is operated more efficiently (Shabbir & Evangelou, 2019). Such techniques have been proven particularly effective for HEVs (Wang et al., 2019) and EVs with hybrid energy storage systems (Hu et al., 2015). In the past decade, the energy management problem of electrified powertrains has been investigated extensively with a range of different proposed strategies, ranging from rule-based to optimization-based. For example, Wei et al. (2017) designed a fuzzy control strategy for HEVs power management by utilising the rules of torques slip behaviours. In Hung & Wu (2012), the energy management problem of EVs powered by a combination of a lithium battery and a supercapacitor module is studied. The proposed integrated optimisation approach gives a solution of component sizing and energy sources control which is proven to save 6% of the electric energy. However, the energy consumption calculation during the design process of high efficiency electrified powertrains generally assumes the vehicle motion following predefined driving cycles (Chen et al., 2017). Hence, the influence of human driver's behaviour to the energy consumption is not considered in these research studies.

There is also a significant body of work in eco-driving, in which it has been shown that the way a vehicle is driven can significantly affect the overall fuel use and subsequent emissions. Research has estimated that 5–10% of fuel can be saved if drivers pursued a more fuel-efficient, economical, and environmentally friendly driving style (Schall & Mohnen, 2017; Xu et al., 2017). In general, eco-driving behaviours can be studied in different scenarios such as: car-following (Zhao & Sun, 2013), cornering (Han et al., 2011), intersection crossing with traffic lights (Sun et al., 2015), and varying road grade (Kamal et al., 2011). The first two scenarios are considered in detail in the present work.

Ojeda et al. (2017) and Han et al. (2018) proposed speed advisory systems for connected vehicles which are claimed to be able to generate a near-optimal speed profile in terms of energy minimisation while also avoiding collisions. Numerically, the problem is solved using a model predictive control-like approach in which real-time implementation is possible by the analytical solution of a state-constrained optimal control problem (OCP). However, the formulation of this problem assumes that the future speed-profile of the preceding vehicle is known, which is unrealistic. In addition, the speed and acceleration preferences of drivers and passengers when following the reference profile are not considered in the problem. Kamal et al. (Kamal et al., 2013; Kamal & Kawabe, 2015) presented an eco-driving assistance system for a host vehicle that uses mixed information including the state of the preceding vehicle, road gradients, and the state of upcoming traffic signals. A dynamic model is used to predict the behaviour of the leader vehicle, which is claimed to be more realistic than assuming a constant speed or acceleration of the leader vehicle. A multi-objective cost function is used for the OCP formulation, which not only maximises fuel economy but also regulates a safe headway distance. Model predictive control (MPC) is used to drive the vehicle optimally by anticipating upcoming road traffic situations. Based on the predicted future states, road slope information, and the fuel consumption model of the engine, it calculates the optimal vehicle control input to drive the vehicle over a prediction horizon. Others have studied the influence of this kind of speed advisory system at network level. Wan et al. (2016) proposed an analytical solution to a fuel consumption minimisation problem. They evaluate the influence of vehicles with such a system on the traffic network, demonstrating that the system not only improves the fuel economy of the vehicle on which it is installed but also it benefits the surrounding vehicles that do not have this system.

Despite the promising work, these existing eco-driving techniques neglect the preferences and comfort of the human driver to follow the optimised speed profile. In the present work, a driver preference model developed recently in (Fleming et al., 2018b; Allison et al., 2019) is adopted which takes into account the naturalistic driver preferences in terms of speed, vehicle spacing, acceleration, and cornering speed. Previous works on speed advisory systems for EVs typically do not consider driver preferences in the optimisation process, and as discussed in previous chapters, including them may have beneficial effects on the user acceptance of such systems.

Therefore, the contribution of this chapter is: a) a novel speed advisory system for EVs, which blends the driver preference maximisation with the energy minimisation in the framework of optimal control theory, b) the formulation of the proposed speed advisory system as a receding horizon problem with a novel terminal cost, such that it has the ability to incorporate real-time information from the vehicle and surrounding environment, and c) the study of the resulting trade-off of driver preference and energy economy, such that it may be used to improve the user acceptance of the speed advisory system.

SPEED ADVISORY PROBLEM

For a speed advisory system, the goal is to generate a reference speed profile based on available information on the ego-vehicle and its surrounding environment, such that by following the reference profile, the vehicle will meet certain performance criteria, such as reducing energy consumption and guaranteeing vehicle safety constraints. In this paper, the speed advisory problem is formulated under the framework of optimal control theory (Kirk, 2004).

Define the system state and the input as:

$$\mathbf{x} = [x, v, \sigma]^T \quad \mathbf{u} = [a]$$

where x and v are the travelled distance and forward velocity of the ego-vehicle respectively, σ is the battery state of charge (SoC), and a the longitudinal acceleration.

The optimal control problem (OCP) for the speed advisory system is formulated as:

$$\min_{\mathbf{u} \in \mathcal{U}} \quad J[\mathbf{x}, \mathbf{u}] \tag{7.1}$$

$$\text{s.t.} \quad \dot{\mathbf{x}} = \mathbf{f}(\mathbf{x}, \mathbf{u}) \tag{7.2}$$

$$\mathbf{g}(\mathbf{x}, \mathbf{u}) \leq 0 \tag{7.3}$$

$$\text{Boundary conditions on } \mathbf{x}, \mathbf{u} \tag{7.4}$$

where $J[\mathbf{x}, \mathbf{u}]$ is the cost function given by:

$$J[\mathbf{x}, \mathbf{u}] = \int_0^T \left(L_{\text{dsm}}(\mathbf{x}, \mathbf{u}, t) + w L_b(\mathbf{x}, \mathbf{u}, t) \right) dt \tag{7.5}$$

The overall stage cost function consists of two terms, $L_{\text{dsm}}(\mathbf{x}, \mathbf{u}, t)$ and $L_b(\mathbf{x}, \mathbf{u}, t)$, where $L_{\text{dsm}}(\mathbf{x}, \mathbf{u}, t)$ is the stage cost related to driver preferences, which penalises certain "unpleasant" driving behaviours, and $L_b(\mathbf{x}, \mathbf{u}, t)$ is the stage cost related to

VEHICLE MOTION AND DRIVER MODELLING

Vehicle Motion Dynamics

Vehicle motion is described by the differential equations:

$$\mathbf{f}_1 : \quad \dot{x} = v \qquad (7.6)$$

$$\mathbf{f}_2 : \quad \dot{v} = a \qquad (7.7)$$

where the speed and acceleration should be constrained as:

$$\mathbf{g}_1 : \quad 0 \leq v \leq v_{max}$$

$$\mathbf{g}_2 : \quad a_{min} \leq a \leq a_{max}$$

and where v_{max} is the maximum vehicle speed, a_{min} is the maximum deceleration, and a_{max} is the maximum acceleration.

Driver Preference Model

In our previous work (Fleming et al., 2018a), an optimal control based driver behaviour model is briefly presented, which includes a cost function to indicate driver preferences and, when minimised, leads to naturalistic driving behaviour.

If the travelled distance profile of the leader vehicle is given by a function $x_L(t)$, where the time t is within an interval $[0,T]$, the headway distance between the ego-vehicle and the leader vehicle is calculated as:

$$s = x_L(t) - x \qquad (7.8)$$

In order to avoid collisions, the following constraint is introduced:

$$\mathbf{g}_3 : \quad s \geq 0$$

The driver model presented in Fleming et al. (2018a) is based on the intelligent driver model (IDM) of Treiber et al. (2000), and has the stage cost function:

$$L_{\text{dsm}}(\mathbf{x},\mathbf{u},t) = \frac{\delta}{v_d}(v - v_d)^2 + \frac{(1 - s/s_d)^2}{(s/s_d)^2 + 1} + a^2 \qquad (7.9)$$

where v_d is the desired vehicle speed, which is the speed the driver will take when they are unconstrained by corners, speed limits, and other vehicles. Minimising this first term $\frac{\delta}{v_d}(v-v_d)^2$ represents the urge to reduce travel time by driving close to this speed. The parameter δ is a parameter which is same as the acceleration exponent in the IDM (Treiber et al., 2000). The second term $\frac{(1-s/s_d)^2}{(s/s_d)^2+1}$ is a penalty function that becomes large for small headway distances, but tends to a constant value as the headway distance becomes large to represent the driver's indifference to far away vehicles. This penalty function reaches its minimum when the headway distance is equal to a desired headway distance s_d which is scaled according to the current speed and defined as $s_d = s_{d,0} + v\tau$. Hence, $s_{d,0}$ is the desired headway distance when the vehicle is stationary, and τ is the desired minimum time gap between vehicles. The final term a^2 in the stage cost L_dsm penalises aggressive acceleration and braking behaviour, in order to improve driver comfort.

In addition, to model driver behaviour when the ego-vehicle is driving through a curved road section, a cornering driver model developed by Reymond et al. (2001) is also incorporated by including the following inequality constraint on the lateral acceleration a_y

$$\mathbf{g_4}: \quad a_y = \kappa(x)v^2 \leq \Gamma_{max} - \Delta v_e^2 \tag{7.10}$$

wher $\kappa(x)$ is the curvature, which is function of the travelled distance, Γ_{max} is maximum lateral acceleration, and Δ is the driver's "safety margin" of error in their estimate of the curvature when approaching the corner. Rearranging (7.10), the constraint g_4 becomes a speed constraint on v:

$$v \leq \sqrt{\frac{\Gamma_{max}}{\kappa(x)+\Delta}} \tag{7.11}$$

Based on the naturalistic driving data collected using automotive data acquisition module (ADAM) (Yan et al., 2017a) as part of the green adaptive control for interconnected vehicles (G-ACTIVE) project (https://g-active.uk/) at the University of Southampton, sensible values of Γ_{max} are about $5-6\ m/s^2$ while Δ should be around $4-5\ rad/km$. We recommend the reader refers to the paper (Fleming et al., 2018b) for more details.

ELECTRIC POWERTRAIN ENERGY CONSUMPTION MODEL

In this chapter, a front-wheel driven EV is modelled by considering the major energy losses in the powertrain and during driving as shown in Figure 7.1. The EV parameters are based on the specification of the Nissan Leaf, which uses a 80 kW and 280 Nm front-mounted synchronous electric motor (Sato et al., 2011). As shown in

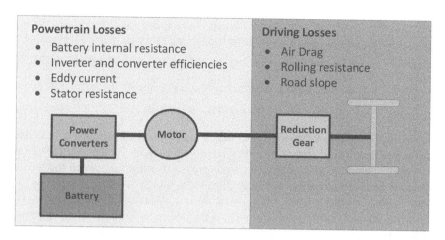

FIGURE 7.1 Electric vehicle structure and losses considered.

Figure 7.1, the EV powertrain contains a battery pack, which powers the motor via DC-DC converters and DC-AC inverters. The motor is coupled to the wheel via a fixed reduction gear.

Driving Losses

The overall driving force F on the wheels can be calculated as (Yan et al., 2018)

$$F = ma + \frac{1}{2}\rho A_f C_d v^2 + C_{rr} mg + mg\sin\theta \qquad (7.12)$$

where m is the vehicle mass, ρ the air density, A_f the frontal area, C_d the coefficient of drag, C_{rr} the coefficient of rolling resistance, and θ is the road gradient. In this calculation, driving losses due to air resistance, rolling resistance and road slope have been taken into account.

The driving force F may be separated as a traction force F_t and braking force F_b which correspond to the positive part and negative part of F, respectively. Their relationship can be expressed as:

$$F = F_t - F_b \qquad (7.13)$$

While the traction force F_t is solely provided by the electric motor, the braking force F_b is the sum of the regenerative braking force F_{bmg} from the motor and the hydraulic braking force F_{bh}. The calculation of the regenerative braking force F_{bmg} is not trivial task because it must consider both the braking force distribution and motor limitations. This issue was generally neglected in the literature by simply assuming $F_{bmg} = \alpha F_b$ with α equal to either one or a fixed proportion (Kuriyama et al., 2010; Vatanparvar et al., 2015).

Receding Horizon Eco-Driving Assistance Systems for Electric Vehicles

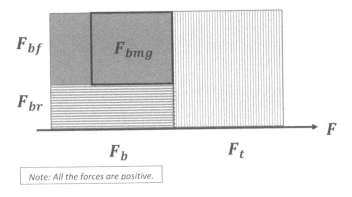

FIGURE 7.2 Forces relationship.

As shown in Figure 7.2, the distribution of the driving force F is considered in several parts. Firstly, the braking force F_b is distributed to front and rear wheels as F_{bf} and F_{br} respectively. For the front wheel driven powertrain considered here, the regenerative braking force F_{bmg} is a fraction of front wheel braking force F_{bf} only.

In order to provide maximum directional stability and the full use of the adhesion condition in the braking process, it is assumed that the front and rear wheel braking forces always meet the ideal braking force allocation curve (Fajri et al., 2016). Hence, the front wheel braking force F_{bf} can be calculated as:

$$F_{bf} = \frac{L - L_a}{L} F_b + \frac{h_g}{mgL} F_b^2 \qquad (7.14)$$

where L is the wheelbase, h_g is the gravity centre height of the vehicle, L_a is the length from the vehicle's center of gravity to the front axle.

Since the front axle is connected to the electric motor, it is possible and convenient to use the electric motor and recover energy rather than the hydraulic brake and dissipate it. However, regenerative braking capabilities are limited by several factors. Firstly, the (negative) minimum torque of the motor $T_{m,min}$ limits the maximum regenerative braking force $F_{mg,max}$ which the motor can provide to the wheel

$$F_{mg,max} = \frac{|T_{m,min}| N_r}{r_w} \qquad (7.15)$$

where N_r is the reduction gear ratio and r_w is the radius of the wheels. Figure 7.3 shows the relationship between front wheel braking force F_{bf} and overall braking force F_b based on the equation (7.14). In addition, the green dash line implies the fraction of regenerative braking force F_{bmg} regarding to F_{bf}, while red solid line represents the amount of hydraulic braking if the regenerative braking cannot meet the requirement of front wheel braking.

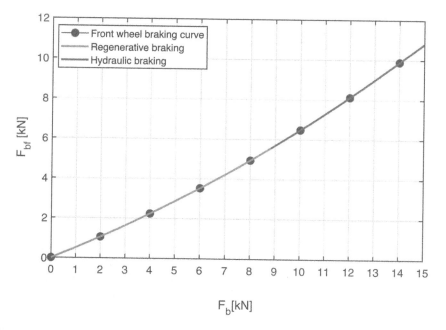

FIGURE 7.3 Front wheel braking curve.

The regenerative braking is also limited by the inability of the electric motor to operate as a generator and charge the battery at low speeds. Fajri et al. demonstrated in (Fajri et al., 2016) that there is a low motor speed threshold below which current is drawn from the battery instead of charging the battery even though motor torque is negative. A switching function modelling the motor speed threshold is defined as:

$$\nu(\omega_m - \omega_{th}) = \begin{cases} 0 & \omega_m - \omega_{th} \leq 0 \\ 1 & \omega_m - \omega_{th} > 0 \end{cases} \tag{7.16}$$

where ω_{th} is the motor speed threshold. This indicator function is equal to zero when the motor speed is lower than the threshold and is equal to one when the motor speed is over the threshold. By combining the above two limitations, the regenerative braking force can be calculated as:

$$F_{bmg} = \min(F_{bf}, F_{rg,max})\nu(\omega_m - \omega_{th}) \tag{7.17}$$

Summarising the equation (7.14) and (7.17), the force from the motor at the wheel can be calculated as:

$$F_m = F_t - F_{bmg} = F_t - \Gamma(F_b) \tag{7.18}$$

where $\Gamma(F_b)$ represents the calculation of the regenerative braking force using equations (7.14, 7.16, 7.17).

POWERTRAIN LOSSES

The electrical system of the vehicle is implemented as a lookup table as shown in Figure 7.4, which maps values of motor torque demand T_m and motor speed ω_m to current consumption values i_b at the battery (Yan et al., 2017b). This map of battery current consumption is calculated by taking into account energy losses due to stator resistance, eddy current losses, inverter and converter efficiencies, and battery internal resistance. Note that the motor torque and resulting battery current on this map may be negative to allow for regenerative braking. As the rate of change of the battery charge Q_b is equal to the battery current i_b, the dynamics of Q_b are modelled as follow by interpolating the 2D lookup table shown in Figure 7.4.

$$\dot{Q}_b = i_b(\omega_m, T_m) \tag{7.19}$$

In this paper, the battery state of charge (SoC) $\sigma(t)$ is considered as the only system state of the electric powertrain. Hence the dynamics of battery SoC $\sigma(t)$ are:

$$\mathbf{f}_3: \quad \dot{\sigma}(t) = \frac{-\dot{Q}_b}{Q_{max}} = \frac{-i_b(\omega_m, T_m)}{Q_{max}} \tag{7.20}$$

where Q_{max} is the maximum battery charge. Equation $\mathbf{f}_3(20)$, together with dynamic equations $\mathbf{f}_1(6)$ and $\mathbf{f}_2(7)$, complete the state space model (7.2). For electric powertrains, energy economy can be evaluated by the change in battery SoC $\Delta\sigma$

$$\Delta\sigma = \sigma(0) - \sigma(T) \tag{7.21}$$

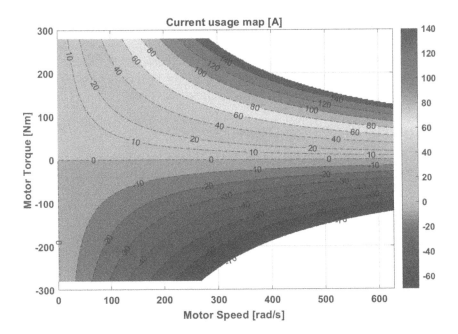

FIGURE 7.4 Battery current usage map of electric powertrain.

Based on the dynamics of the battery SoC, minimising $\Delta\sigma$ is equivalent to minimising the integral of the battery current i_b. Therefore, for the proposed economical speed advisory system, the stage cost related to the powertrain energy economy is defined as the minimisation of the battery current i_b.

$$L_b = i_b(\omega_m, T_m) \tag{7.22}$$

The motor crankshaft and the wheel shaft are connected via a fixed reduction gear with gear ratio N_r so that the motor torque and crankshaft speed can be calculated as:

$$\omega_m = \frac{N_r}{r_w} v \tag{7.23}$$

$$T_m = \frac{r_w}{N_r} F_m \tag{7.24}$$

In addition, the motor power and torque are constrained as:

$$\mathbf{g}_5: \quad |T_m \omega_m| \leq P_{max}$$

$$\mathbf{g}_6: \quad T_{m,min} \leq T_m \leq T_{m,max}$$

In the presented case study, P_{max} is the maximum power of the motor which is equal to $80\,kW$, $T_{m,max}$ is the maximum torque which is equal to $280\,Nm$, $T_{m,min} = -280\,Nm$ is the minimum torque.

In conclusion, for the speed advisory OCP problem, the system dynamics $\dot{\mathbf{x}} = f(\mathbf{x}, \mathbf{u})$ include three equations: $\mathbf{f}_1$, $\mathbf{f}_2$, and $\mathbf{f}_3$. The inequality constraints considered in this paper are summarised as $\mathbf{g}_1$, $\mathbf{g}_2$, $\mathbf{g}_3$, $\mathbf{g}_4$, $\mathbf{g}_5$, and $\mathbf{g}_6$.

FULL-HORIZON OPTIMISATION

The OCP (7.1) described so far is a full horizon problem, which means that the system dynamics (7.2), inequality constraints (7.3), and cost function (7.5) need to be evaluated for the whole horizon $[0, T]$.

Boundary Conditions

For the full-horizon optimisation, the boundary condition for the travelled distance of the ego-vehicle is defined as:

$$x(0) = x_0 \tag{7.25}$$

$$x(T) = x_0 + v_m T \tag{7.26}$$

Receding Horizon Eco-Driving Assistance Systems for Electric Vehicles

where x_0 is the initial distance of the ego-vehicle. The average travel speed v_m and journey time T can be set according to different driving missions. In addition, it is assumed the kinetic energy of the ego-vehicle is same before and after the journey by fixing the initial and final vehicle speed to be equal.

$$v(0) = v(T) \tag{7.27}$$

Finally, for the battery SoC $\sigma(t)$, the initial condition is set, while the final condition is free to allow for solutions with different levels of energy economy.

$$\sigma(0) = 0.7 \tag{7.28}$$

$$\sigma(T) = free \tag{7.29}$$

This problem formulation has the benefit of being unified for two typical driving scenarios: car-following and cornering. For the following two case studies, the variables in boundary conditions (7.25–7.27) are defined as follows:

$$x_0 = 0 \quad v_m = 10 \quad T = 60$$

CAR-FOLLOWING CASE

In this section, the ego-vehicle is assumed to be following a leader vehicle on a straight road. The leader vehicle travels $600m$ in $60s$ with the speed profile shown in Figure 7.5 by the purple dashed line, which has sections of acceleration, deceleration, and constant speed. The initial travelled distance of the leader vehicle is set as:

$$x_L(0) = 10,$$

which means that the leader vehicle is initially $10m$ in front of the ego-vehicle.

The speed optimisation problem is solved with the full-horizon cost (7.5), giving three solutions which are shown in Figure 7.5. The blue line represents the solution when $w = 0$, which means that it only minimises the driver model cost L_{dsm}. The green line represents the solution when $w = 1e6$, which is the situation that the energy economy cost L_b has a strong weighting. As shown in the figure, for the solution with the driver model cost only, the ego-vehicle speed profile mimics the leader profile with some delay, such that the headway distance is maintained close to the desired headway distance s_d. For the solution with a large w, the vehicle shows a "pulse-and-glide" type speed profile, which results in a high final SoC and better energy economy while ignoring the driver preferences. Most importantly, the headway distance $s(t)$ is changing dramatically. The ego-vehicle firstly drives very close to the leader vehicle which is a dangerous manoeuvre. After around $25s$, the

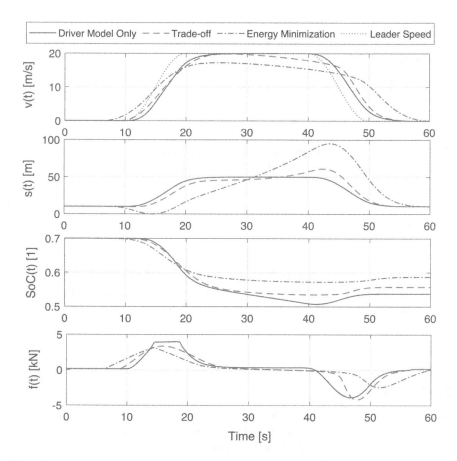

FIGURE 7.5 Full-horizon solutions under different w [car-following case].

ego-vehicle starts to coast down which results in a large headway distance to the leader vehicle. Therefore, even though the green speed profile provides the best energy efficiency, it is unacceptable for a human driver.

In order to investigate the compromise between driver preference and energy economy, the full-horizon OCP is solved under a series of different weighting factors w from 0 to 1e6. For each weighting factor w, the deviation of battery state of charge ΔSoC and the mean driver model cost $1/T \int_0^T L_{\text{dsm}}\, dt$ is plotted in Figure 7.6. In this way, the Pareto curve shown in the figure demonstrates the trade-off behaviour of the speed optimisation problem between energy economy and driver preference. In addition, one solution marked with a solid red spot in Figure 7.6 is selected as a trade-off solution, which is also plotted in Figure 7.5 in red. The trade-off solution incorporates the driver preferences in terms of naturalistic headway distances, which is similar to the driver only case (blue profile). On the other hand, the energy consumption of this solution is reduced by a large amount compared to the driver model only solution (blue profile).

Receding Horizon Eco-Driving Assistance Systems for Electric Vehicles 155

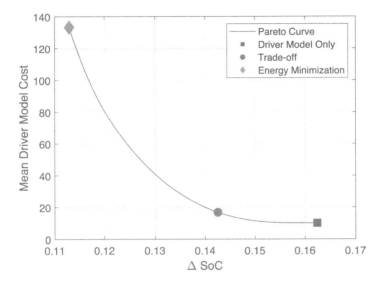

FIGURE 7.6 Pareto curve under different w [car-following case].

CORNERING CASE

Next, the optimal control problem is solved assuming that the ego-vehicle is driving along a fixed length road of varying road curvature. The simple curvature profile used for the $1200m$ route is shown in Figure 7.7, which contains two cornering sections ($250m$ to $1/T \int_0^T L_{dsm}\, dt$ and $850m$ to $950m$). The solution is found for the section between the two corners from $300m$ to $900m$ as shown in red in Figure 7.7, and it is assumed that there is no leader vehicle in front of the ego-vehicle so that the road geometry alone determines the driver behaviour. The initial and final speed of the ego-vehicle is set as: cornering behaviour is modelled in the driver preference model by the inequality constraint (7.10). As demonstrated in the previous subsection, energy reduction is achieved by choosing different headway distance profiles with a fixed average speed. However, for cornering situations, where headway distance is unimportant, energy consumption is mainly influenced by the choice of average speed. Hence, for the full-horizon optimal control, the average speed v_m is set free. And the variables in boundary conditions are defined as follow:

$$v(0) = v(T) = 4.5$$

Similar to the car-following case, three solutions for different weighting factors w are presented in Figure 7.8. The blue line represents the solution when $w = 0$, which means that it only minimises the driver model cost L_{dsm} and represents the natural behaviour of a human driver. The solution marked by the green line represents the solution when $w = 1e6$, which is the situation that the energy consumption is minimised. For this solution, the ego-vehicle is driven in a pulse-and-glide manner. And the red line solution is a trade-off case with $w = 1e3$. In the velocity subplot in

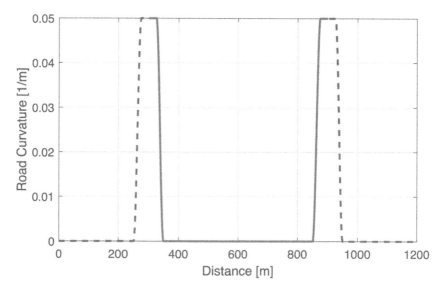

FIGURE 7.7 A case of road curvature.

Figure 7.8, the purple dashed line is the cornering speed constraint as described in (7.11). As it can be seen, the speed profiles of the three solutions all satisfy the constraints on lateral acceleration when negotiating the curve. However, the difference in energy consumption of the three solutions is relatively small compared to the car-following case. This suggests that the pulse-and-glide behaviour (green solution) for curve driving is not as beneficial to reduce energy consumption as it is in the car-following case.

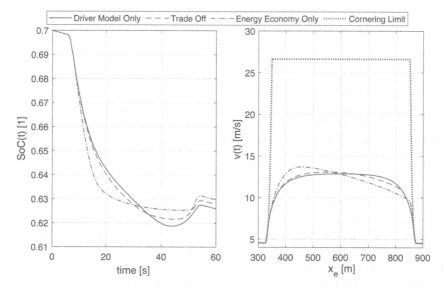

FIGURE 7.8 Trade-off solutions under different w [cornering case].

RECEDING HORIZON CONTROL

The OCP (7.1) proposed in the previous section is a full-horizon problem. However, it is not possible to solve the OCP (7.1) in a real-time implementation because the future state of the ego-vehicle and surrounding environment, for example the leader vehicle speed profile, is unknown. Therefore, receding horizon control is a strong candidate for implementing the speed advisory system in real-time. In receding horizon control, also known as MPC, the control action is produced by minimising the cost function at each sampling instant using the current state and predicted future state of the system for a shorter finite time horizon. The optimisation yields an optimal control sequence and only the first control in this sequence is applied to the system. The reader may refer to Camacho & Alba (2013) for the practical implementation of the receding horizon control technique.

When moving from the full-horizon optimisation to the receding-horizon optimisation, it is possible to introduce conservatism. Hence, the receding horizon control-based formulation should be designed so that the solution is as close as possible to the full horizon problem to preserve optimality. In this chapter, compared to the full horizon formulation, the performance of the receding horizon control is mainly affected by two factors: the prediction of the future behaviour of the leader vehicle, and the short horizon cost function.

BOUNDARY CONDITIONS

As the boundary conditions for the end time T cannot be imposed in the receding horizon OCP, the boundary conditions are only applied for the start of the optimisation, which can be summarised as:

$$x(0) = x_0 \tag{7.30}$$

$$v(0) = v_0 \tag{7.31}$$

$$\sigma(0) = 0.7 \tag{7.32}$$

LEAD VEHICLE TRAJECTORY PREDICTION

One key benefit of using the novel driver preference model is its ability to maintain a natural and comfortable headway distance to the leader vehicle. However, in practice, the driving behaviour of the leader vehicle for the future is hard to accurately predict. In this chapter, it is assumed that the travelled distance x_L and the speed v_L are measurable at each point in time. Since the receding horizon is relatively small compared to the whole journey time, the leader vehicle is assumed to be travelling at a constant speed during the prediction horizon, defined as t_p. If the present time is defined as t_0, $x_L(t_0)$ and $v_L(t_0)$ are the measured travel distance and speed of the leader vehicle respectively. Therefore, the predicted travelled trajectory during the prediction horizon can be calculated as:

$$x_L(t) = x_L(t_0) + v_L(t_0)t \quad t \in [t_0, t_f] \tag{7.33}$$

where $t_f = t_0 + t_p$ is the end time of the prediction horizon. The headway distance s may be calculated by substituting (7.33) into (7.8).

Receding Horizon Cost Function

The ideal receding horizon cost function can be expressed in the following form, which is equal to the full horizon function.

$$J[\mathbf{x},\mathbf{u}] = \int_0^{t_0} \left(L_{\text{dsm}}(\mathbf{x},\mathbf{u},t) + wL_b(\mathbf{x},\mathbf{u},t)\right) dt$$
$$+ \int_{t_0}^{t_f} \left(L_{\text{dsm}}(\mathbf{x},\mathbf{u},t) + wL_b(\mathbf{x},\mathbf{u},t)\right) dt \qquad (7.34)$$
$$+ \int_{t_f}^{T} \left(L_{\text{dsm}}(\mathbf{x},\mathbf{u},t) + wL_b(\mathbf{x},\mathbf{u},t)\right) dt$$

Here, $\int_0^{t_0} \left(L_{\text{dsm}}(\mathbf{x},\mathbf{u},t) + wL_b(\mathbf{x},\mathbf{u},t)\right) dt$ is the cost related to the driving that has already taken place up to the present time, which is an unchangeable fixed value, and $\int_{t_f}^{T} \left(L_{\text{dsm}}(\mathbf{x},\mathbf{u},t) + wL_b(\mathbf{x},\mathbf{u},t)\right) dt$ corresponds to the cost after the prediction horizon until the end of the journey, which is unknown.

Therefore, the receding horizon cost function could be simplified as:

$$J[\mathbf{x},\mathbf{u}] = \phi(\mathbf{x}(t_f)) + \int_{t_0}^{t_f} \left(L_{\text{dsm}}(\mathbf{x},\mathbf{u},t) + wL_b(\mathbf{x},\mathbf{u},t)\right) dt \qquad (7.35)$$

where $\phi(\mathbf{x}(t_f))$ is a terminal cost such that:

$$\phi(\mathbf{x}(t_f)) = \int_{t_f}^{T} \left(L_{\text{dsm}}(\mathbf{x},\mathbf{u},t) + wL_b(\mathbf{x},\mathbf{u},t)\right) dt \qquad (7.36)$$

However, for receding horizon control, the leader vehicle state beyond the prediction horizon t_f is unknown, which means that the terminal cost $\phi(\mathbf{x}(t_f))$ cannot be calculated exactly by (7.36). Hence, in the case study, different approximate terminal cost functions $\phi(\mathbf{x}(t_f))$ are explored in order to find a suitable candidate which is both numerically friendly and provides a solution that is close to the corresponding part of the full horizon optimal solution. The procedure for solving the receding horizon problem is presented in Algorithm 1.

Algorithm 1 Receding Horizon Control

1. Measure the system state $\mathbf{x}$, the leader vehicle distance x_L, and speed v_L at t_0

Receding Horizon Eco-Driving Assistance Systems for Electric Vehicles

2. Calculate the predicted leader vehicle travel trajectory x_L for the horizon $[t_0, t_f]$
3. Solve the receding horizon control problem for the horizon $[t_0, t_f]$ by GPOPS II
4. Extract the optimal input **u** from solution for the sampling period $[t_0, t_0 + t_s]$
5. Zero hold the input **u** during the sampling period $[t_0, t_0 + t_s]$
6. Apply the input and solving the dynamics during the sampling period $[t_0, t_0 + t_s]$
7. Step forward $t_0 = t_0 + t_s$

TERMINAL COST SELECTION

The performance of the receding horizon approach should be as close to the full horizon formulation as possible. In order to explore the influence of the terminal cost, a simple car-following case study is proposed, where it is assumed that the ego-vehicle is following the leader vehicle on a straight road. The leader vehicle velocity profile is same as the profile used in the full-horizon optimisation. The full journey time T of the leader vehicle is 60s, and the overall travel distance L is equal to 600m.

The ego-vehicle is controlled to follow the leader vehicle and the initial condition of the states and leader vehicle distance are set as:

$$x(0) = 0 \, m \qquad x_L(0) = 0 \, m$$

$$v(0) = 0 \, m/s \qquad \delta(0) = 0.7$$

MPC without Terminal Cost

The first attempt is to solve the combined optimisation problem without any terminal cost, which means:

$$\phi(t_f) = 0 \tag{7.37}$$

The optimisation is solved twice with same two different weighting factors w as in the previous part, which are $w = 0$ and $w = 1e6$. As shown in Figure 7.9 in blue, for the case $w = 0$ the solution is similar to the whole-horizon solution. However, for the case $w = 1e6$, the solution is very different from the whole-horizon case. The travelled distance of the ego-vehicle x is around 350m, which is much less than the leader vehicle (600m). Hence, the ego-vehicle travels less distance in order to reduce the energy consumption. The main reason for this different performance to the whole-horizon case is that the terminal constraint (7.26) is missing for the receding horizon case. In addition, the receding horizon control may be unstable when applied without a terminal cost corresponding to some feasible future control action as suggested in the literature (Mayne et al., 2000). Therefore, in order to complement this shortcoming of the receding horizon control, a proper terminal cost $\phi(t_f)$ needs to be included in the receding horizon optimisation.

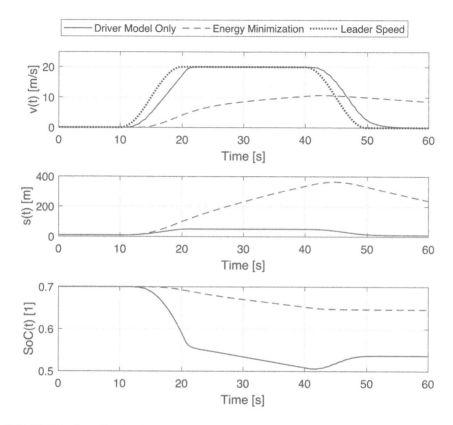

FIGURE 7.9 Receding horizon control solutions under different w [no terminal cost].

MPC with Terminal Cost

In this subsection, a novel terminal cost is defined by constructing a feasible ego-vehicle speed profile during period $[t_f, t_e]$ and evaluating the remaining cost according to this profile. As for the whole-horizon optimisation, the leader vehicle travels a fixed distance L. If the ego-vehicle also travels the same distance L and the average speed v_a of the leader is known, the end time of the optimisation may be calculated as:

$$t_e = \frac{L - x(t_f)}{v_a} + t_f. \tag{7.38}$$

To define the terminal cost, the ego-vehicle speed profile shown in Figure 7.10 is assumed, which has two parts: acceleration or deceleration to the average speed v_a and travelling at the constant speed v_a. The system dynamics during $[t_f, t_e]$ then can be calculated based on this defined speed profile. Here, define the system state and input during $[t_f, t_e]$ as $\mathbf{x}^*$ and $\mathbf{u}^*$ respectively. The energy economy cost L_b

Receding Horizon Eco-Driving Assistance Systems for Electric Vehicles

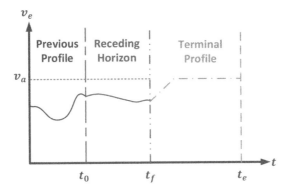

FIGURE 7.10 Speed profile of the remaining journey.

may then be calculated for the interval $[t_f, t_e]$ for this velocity profile. In this way, the terminal cost for time period $[t_f, t_e]$ may be defined as:

$$\phi(\mathbf{x}(t_f)) = \int_{t_f}^{t_e} wL_b(\mathbf{x}^*, \mathbf{u}^*, t) dt \tag{7.39}$$

The result of the receding-horizon optimisation with this terminal cost is shown in Figure 7.11. When $w = 1e6$, the energy consumption decreases compared to $w = 0$ which results in a higher terminal *SoC*. However, compared to the green solution shown in Figure 7.9, the travelled distance of the solutions shown in Figure 7.11 are both 600*m*, which means that the energy reduction here is not due to travelling at a lower average speed. In Table 7.1, the solutions of the receding horizon control are compared quantitatively with the whole horizon control. The overall cost $\int_0^T L_b \, dt$ is divided by the maximum battery charge, which is equal to the deviation of the battery SoC. This value is used to evaluate the energy economy of the solution. In order to quantify how "natural" the driver behaviour is for a specific solution, the average driver model cost $\int_0^T L_{dsm} \, dt/T$ is also calculated.

Firstly, the two control schemes are compared when $w = 0$ so that only the driver model cost L_{dsm} is minimised. As shown in the table, the solution of the whole horizon method gives a lower driver model cost while the energy economy is very similar between the two solutions. The difference between the two solutions may be explained by noting that the receding horizon method does not have knowledge of the future speed profile of the leader vehicle. Hence, the receding horizon control is a little conservative compared to the whole horizon control.

Secondly, the whole horizon and receding horizon methods are compared with $w = 1e6$. In this case, the energy economy cost has a much stronger weighting than the driver model cost. As shown in the table, the energy economy of the whole-horizon method is slightly better than the receding horizon methods and the driver model cost of whole horizon case is much lower than the receding horizon case. By

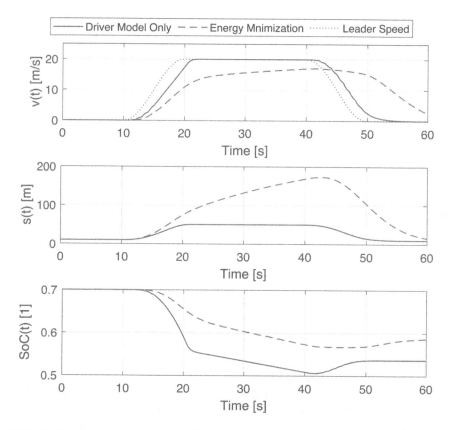

FIGURE 7.11 Receding horizon control solutions under different w [with terminal cost].

comparing the green speed profile in Figure 7.5 and Figure 7.11, it can be seen that for the whole horizon control, the ego-vehicle may accelerate before the leader vehicle in order to produce a pulse-and-glide behaviour which results in better energy economy. However, for receding horizon control, the response of the ego-vehicle is delayed compared to the leader vehicle because the future behaviour of the leader vehicle

TABLE 7.1

Energy Economy Compared between Whole Horizon and Receding Horizon Solutions

	Horizon	$\int_0^T L_b\, dt/Q_{max}$	$\int_0^T L_{dsm}\, dt/T$	Distance
$w = 0$	Whole	0.0484	9.99	600m
	Receding	0.0475	12.8	600m
$w = 1e6$	Whole	0.0337	73.2	600m
	Receding	0.0341	137	600m

is unknown. For example, for the start-up behaviour, in order to avoid collision, the ego-vehicle will only start speeding after detecting the leader vehicle already do so.

What's more, with the proposed terminal cost, the travelled distance of the ego-vehicle is the same as the whole-horizon case for different weighting factors, which means that the terminal constraint (7.26) is satisfied, and the terminal cost leads to similar results compared to the whole-horizon stage cost. Even though the performance of the receding control with the proposed terminal cost is not exactly the same as in the optimal whole horizon case, it is acceptable considering the benefit of being real-time implementable.

TEST CASE UNDER REAL-WORLD DRIVING DATA

In order to further validate the performance of the proposed speed advisory system, in this section, a real-world car-following test case is simulated. A set of naturalistic driving data collected by ADAM (Yan et al., 2017a) as part of the G-ACTIVE project is used. ADAM is a data collection device placed in a study participant's own vehicle, which collects data on vehicle position and velocity using GPS as well as the headway distance to the leader vehicle by post-processing of recorded stereo video.

The OCP problem (7.1) is solved assuming the leader speed profile is following one set of driving data collected by ADAM as the black dashed line shown in Figure 7.12. The duration of this case study is $700s$ and the start position of the lead car is assumed to be $3m$ in front of the ego-vehicle. The leader vehicle is assumed to have the same

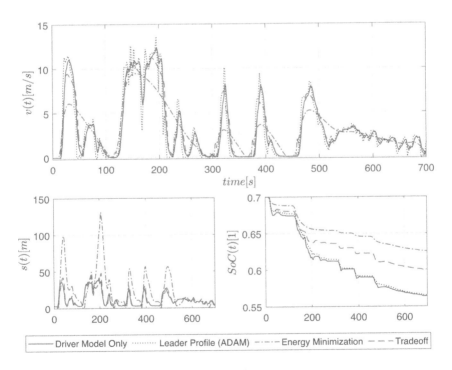

FIGURE 7.12 Trade-off solutions with ADAM data.

TABLE 7.2
Energy Saving Comparison

Solutions	ΔSoC	Energy Saving
Leader Vehicle (Benchmark)	0.1350	0%
Driver Model Only (Red)	0.1363	−0.9% ↓
Energy Minimisation (Blue)	0.0748	44.6% ↑
Trade-off (Green)	0.1003	25.7% ↑

powertrain as the ego one. Hence, the SoC of the leader vehicle is also plotted in the same SoC subplot with black dashed line. Table 7.2 shows the energy consumption of three different solutions compared to the leader vehicle. In the following, these three solutions are discussed in detail.

The OCP problem is first solved in the full-horizon framework with a weighting factor $w = 1e6$ which corresponds to the least energy consumption case. The solution is shown in Figure 7.12 in blue. It is clear that the solution demonstrates the pulse-and-glide behaviour which has already been identified as the most energy-efficient driving behaviour in the existing literature (Li & Peng, 2012). However, this pulse-and-glide is difficult to implement for human drivers because the required timing of the acceleration and coasting down is difficult to determine. In addition, as shown in the headway profile $s(t)$ in Figure 7.12, the solution contains several moments of near-zero headway distance and several large peaks. The near-zero headway distance scenarios should be definitely avoided for safety reasons, and the large headway distance scenarios are also not preferred by human drivers. Hence, even though this solution results in the least battery charge consumption, it is not likely to be natural enough to satisfy a human driver.

The OCP problem is then solved in the receding horizon control framework proposed in this section with the novel terminal cost (7.39). The solution shown in red corresponds is the solution with $w = 0$. This demonstrates the driving behaviour proposed by the novel driver preference model. As one can see from the speed profile and headway profile, the ego-driver is essentially mimicking the behaviour of the leader vehicle and maintains the desired headway distance s_d. As shown in the SoC subplot, the energy consumption of this solution is almost the same as the leader vehicle. However, compared to the least energy consumption case, the energy consumption is almost doubled.

The main contribution of the proposed speed advisory system is the ability to make a compromise between the least energy consumption case (blue) and the driver model only case (red). The solution marked by a green colour corresponds to the solution with $w = 1e3$ in the receding horizon control framework, which is a trade-off of the previous two cases. As shown in Figure 7.12, the speed and headway profiles for this solution are close to the driver model only case, which means that human drivers will find the profiles easier and more natural to follow than the least energy consumption case. In terms of energy consumption, this solution uses around 25.7% less electrical energy compared to the leader vehicle.

REFERENCES

Achour, H., Carton, J., & Olabi, A. G. (2011). Estimating vehicle emissions from road transport, case study: Dublin city. *Applied Energy*, 88(5), 1957–1964.

Af Wåhlberg, A. (2006). Short-term effects of training in economical driving: Passenger comfort and driver acceleration behavior. *International Journal of Industrial Ergonomics*, 36(2), 151–163.

Af Wåhlberg, A. E. (2007). Long-term effects of training in economical driving: Fuel consumption, accidents, driver acceleration behavior and technical feedback. *International Journal of Industrial Ergonomics*, 37(4), 333–343.

Allison, C., Fleming, J., Yan, X., Stanton, N., & Lot, R. (2019). From the simulator to the road—realization of an in-vehicle interface to support fuel-efficient eco-driving. In: *International Conference on Intelligent Human Systems Integration*, pages 814–819. Springer.

Camacho, E. F., & Alba, C. B. (2013). *Model predictive control*. Springer science & business media.

Castaings, A., Lhomme, W., Trigui, R., & Bouscayrol, A. (2016). Comparison of energy management strategies of a battery/supercapacitors system for electric vehicle under real-time constraints. *Applied Energy*, 163, 190–200.

Chen, F., Taylor, N., & Kringos, N. (2015). Electrification of roads: Opportunities and challenges. *Applied Energy*, 150, 109–119.

Chen, Z., Xiong, R., Wang, C., & Cao, J. (2017). An on-line predictive energy management strategy for plug-in hybrid electric vehicles to counter the uncertain prediction of the driving cycle. *Applied Energy*, 185, 1663–1672.

Fajri, P., Lee, S., Prabhala, V. A. K., & Ferdowsi, M. (2016). Modeling and integration of electric vehicle regenerative and friction braking for motor/dynamometer test bench emulation. *IEEE Transactions on Vehicular Technology*, 65(6), 4264–4273.

Fleming, J., Yan, X., Allison, C., Stanton, N., & Lot, R. (2018a). Driver modeling and implementation of a fuel-saving ADAS. In: *2018 IEEE International Conference on Systems, Man, and Cybernetics (SMC)*, pages 1233–1238. IEEE.

Fleming, J. M., Allison, C. K., Yan, X., Lot, R., & Stanton, N. A. (2019). Adaptive driver modelling in ADAS to improve user acceptance: A study using naturalistic data. *Safety Science*, 119, 76–83.

Han, J., Park, Y., & Park, Y. S. (2011). Adaptive regenerative braking control in severe cornering for guaranteeing the vehicle stability of fuel cell hybrid electric vehicle. In: *2011 IEEE Vehicle Power and Propulsion Conference*, pages 1–5. IEEE.

Han, J., Sciarretta, A., Ojeda, L. L., De Nunzio, G., & Thibault, L. (2018). Safe-and eco-driving control for connected and automated electric vehicles using analytical state-constrained optimal solution. *IEEE Transactions on Intelligent Vehicles*, 3(2), 163–172.

Hu, S., Liang, Z., & He, X. (2015). Ultracapacitor-battery hybrid energy storage system based on the asymmetric bidirectional z-source topology for EV. *IEEE Transactions on Power Electronics*, 31(11), 7489–7498.

Hung, Y. H., & Wu, C. H. (2012). An integrated optimization approach for a hybrid energy system in electric vehicles. *Applied Energy*, 98, 479–490.

Kamal, M. A. S., & Kawabe, T. (2015). Eco-driving using real-time optimization. In: *Control Conference (ECC), 2015 European*, pages 111–116. IEEE.

Kamal, M. A. S., Mukai, M., Murata, J., & Kawabe, T. (2011). Ecological vehicle control on roads with up-down slopes. *IEEE Transactions on Intelligent Transportation Systems*, 12(3), 783–794.

Kamal, M. A. S., Mukai, M., Murata, J., & Kawabe, T. (2013). Model predictive control of vehicles on urban roads for improved fuel economy. *IEEE Transactions on Control Systems Technology*, 21(3), 831–841.

Ke, W., Zhang, S., He, X., Wu, Y., & Hao, J. (2017). Well-to-wheels energy consumption and emissions of electric vehicles: Mid-term implications from real-world features and air pollution control progress. *Applied Energy*, *188*, 367–377.

Kirk, D. E. (2004). *Optimal control theory: an introduction*. Courier Corporation. Chelmsford, Massachusetts, United States.

Kuriyama, M., Yamamoto, S., & Miyatake, M. (2010). Theoretical study on eco-driving technique for an electric vehicle with dynamic programming. In: *2010 International Conference on Electrical Machines and Systems*, pages 2026–2030. IEEE.

Li, S. E., & Peng, H. (2012). Strategies to minimise the fuel consumption of passenger cars during car-following scenarios. *Proceedings of the Institution of Mechanical Engineers, Part D: Journal of Automobile Engineering*, *226*(3), 419–429.

Mayne, D. Q., Rawlings, J. B., Rao, C. V., & Scokaert, P. O. (2000). Constrained model predictive control: Stability and optimality. *Automatica*, *36*(6), 789–814.

Ojeda, L. L., Han, J., Sciarretta, A., De Nunzio, G., & Thibault, L. (2017). A real-time eco-driving strategy for automated electric vehicles. In: *Decision and Control (CDC), 2017 IEEE 56th Annual Conference on*, pages 2768–2774. IEEE.

Reymond, G., Kemeny, A., Droulez, J., & Berthoz, A. (2001). Role of lateral acceleration in curve driving: Driver model and experiments on a real vehicle and a driving simulator. *Human Factors*, *43*(3), 483–495.

Sato, Y., Ishikawa, S., Okubo, T., Abe, M., & Tamai, K. (2011). Development of high response motor and inverter system for the nissan leaf electric vehicle. Technical report, SAE Technical Paper.

Schall, D. L., & Mohnen, A. (2017). Incentivizing energy-efficient behavior at work: An empirical investigation using a natural field experiment on eco-driving. *Applied Energy*, *185*, 1757–1768.

Shabbir, W., & Evangelou, S. A. (2019). Threshold-changing control strategy for series hybrid electric vehicles. *Applied Energy*, *235*, 761–775.

Sun, C., Moura, S. J., Hu, X., Hedrick, J. K., & Sun, F. (2015). Dynamic traffic feedback data enabled energy management in plug-in hybrid electric vehicles. *IEEE Transactions on Control Systems Technology*, *23*(3), 1075–1086.

Treiber, M., Hennecke, A., & Helbing, D. (2000). Congested traffic states in empirical observations and microscopic simulations. *Physical Review E*, *62*(2), 1805.

Vatanparvar, K., Wan, J., & Al Faruque, M. A. (2015). Battery-aware energy-optimal electric vehicle driving management. In: *2015 IEEE/ACM International Symposium on Low Power Electronics and Design (ISLPED)*, pages 353–358. IEEE.

Wan, N., Vahidi, A., & Luckow, A. (2016). Optimal speed advisory for connected vehicles in arterial roads and the impact on mixed traffic. *Transportation Research Part C: Emerging Technologies*, *69*, 548–563.

Wang, F., Zhang, J., Xu, X., Cai, Y., Zhou, Z., & Sun, X. (2019). A comprehensive dynamic efficiency-enhanced energy management strategy for plug-in hybrid electric vehicles. *Applied Energy*, *247*, 657–669.

Wei, Z., Xu, J., & Halim, D. (2017). Hev power management control strategy for urban driving. *Applied Energy*, *194*, 705–714.

Xu, Y., Li, H., Liu, H., Rodgers, M. O., & Guensler, R. L. (2017). Eco-driving for transit: An effective strategy to conserve fuel and emissions. *Applied Energy*, *194*, 784–797.

Yan, X., Fleming, J., Allison, C., & Lot, R. (2017a). Portable automobile data acquisition module (ADAM) for naturalistic driving study. In: *Proceedings of the 15th European Automotive Congress*.

Yan, X., Fleming, J., & Lot, R. (2017b). Modelling and energy management of parallel hybrid electric vehicle with air conditioning system. In: *2017 IEEE Vehicle Power and Propulsion Conference (VPPC)*, pages 1–5, Dec. DOI: 10.1109/VPPC.2017.8330923.

Yan, X., Fleming, J., & Lot, R. (2018). Fuel economy and naturalistic driving for passenger road vehicles. In: *2018 IEEE Vehicle Power and Propulsion Conference (VPPC)*, pages 1–6. IEEE.

Zhao, L., & Sun, J. (2013). Simulation framework for vehicle platooning and car-following behaviors under connected-vehicle environment. *Procedia-Social and Behavioral Sciences*, *96*, 914–924.

8 In Simulator Assessment of a Feedforward Visual Interface to Reduce Fuel Use

INTRODUCTION

The International Energy Agency estimate that transportation accounts for 35% of overall global energy use (IEA, Key world energy statistics, 2017), with passenger cars and light-duty vehicles alone accounting for 21% of global energy use and 24% of global carbon dioxide (CO_2) emissions, contributing to yearly increases in atmospheric CO_2. The primary concern relating to the rising levels of atmospheric CO_2 is due to its position as a greenhouse gas, causing global warming via an increase in radiative forcing of Earth's surface. These changes are predicted to have large-scale negative economic, social, and ecological impacts across multiple domains, including agriculture (Fischer et al., 2005), biodiversity (Botkin et al., 2007), and human health (McMichael et al., 2006). Within this context, it is estimated that 14% of global mean temperature change will be a direct consequence of transportation, with automobiles being the primary contributor (Skeie et al., 2009).

Approaches to addressing the problem of road vehicle emissions currently being explored include the promotion of alternative transportation schemes such as cycle to work events (Rose & Marfurt, 2007), development of alternative, environmentally friendly, and fuel-efficient vehicular drivetrains (Piatkowski et al., 2014), for example the use of hybrid, electric, or hydrogen fuel cell vehicles (Lorf et al., 2013) and the use of cleaner alternative fuel sources such as compressed natural gas (Windecker & Ruder, 2013). For many, however, the initial upfront cost of such alternatives and the lack of a dedicated infrastructure (Yilmaz & Krein, 2013) to support a shift in vehicle type mean that they are not viable solutions in the immediate future.

Although some CO_2 emission is unavoidable from the use of vehicles with internal combustion engines, multiple factors influence the quantity of fuel burned and the subsequent quantity of emissions. In addition to vehicular drivetrains (Chan, 2007) and mechanical systems (Vining, 2009), driving style is a key determinant of overall fuel usage (McIlroy et al., 2013). Barkenbus (2010) proposed that altering the way a vehicle is driven could noticeably reduce both fuel use and emissions. Should individuals adopt a measured driving style, referred to as eco-driving, typified by behaviours such as gentle acceleration, early gear changes, limiting the engine to approximately 2,500 revolutions per minute (RPM), anticipating traffic flow to minimise braking, driving below the speed limit, and limiting unnecessary idling, overall fuel use and emissions could be considerably reduced (Barkenbus, 2010). Eco-driving, the behavioural approach to emissions reduction, has been demonstrated to hold considerable potential, with previous research suggesting that overall fuel usage is reduced by approximately 5%–10% (Martin et al., 2012). This is true for both manual transmission vehicles and automatic ones where control over gear changes is handled by the vehicle and not the driver (Larue et al., 2014). Importantly, these behaviours are not centered purely on the idea that drivers should slow down, with previous research indicating that engagement in eco-driving can lead to notable fuel savings and emission reduction without dramatically increasing overall journey time (Birrell et al., 2013).

Whilst eco-driving is, from the drivers' perspective, a cost-effective way to reduce fuel use (Birol, 2011), requiring no direct financial investment (Saboohi & Farzaneh, 2009), drivers are required to invest a considerable amount of time and effort into developing and maintaining eco-driving skills. Previous research has demonstrated that some behaviours, including minimisation of RPM and early gear changes, are difficult to habitualise and engage with (Delhomme et al., 2013). Indeed, past studies have identified that the practice of eco-driving is associated with a noticeable increase in workload (Pampel et al., 2015). Drivers may simply feel that the effort required to initially learn, engage, and maintain eco-driving behaviours are simply not worth it for the limited financial saving (Dogan et al., 2014). Although fuel economy is of increasing importance to drivers (Turrentine & Kurani, 2007), especially due to rising fuel prices (Heyes et al., 2015), for many the estimated £5 a week saving (Beusen et al., 2009) is not viewed as a worthwhile investment for the level of required effort. One approach in supporting drivers to eco-drive is via the use of specially designed in-vehicle interfaces (Tulusan et al., 2012). The provision of feedback based on current and previous actions can encourage, support, and reinforce eco-driving practices (Froehlich et al., 2009; Meschtscherjakov et al., 2009). Numerous studies, both in simulators and real-world driving, have demonstrated the potential benefit of eco-feedback devices (Young et al., 2009; Boriboonsomsin et al., 2010; Mensing et al., 2014).

One flaw common to eco-driving feedback devices, however, is they typically do not adapt to the present driving context, instead only providing feedback on the fuel efficiency of drivers' past actions, which may be detrimental to the performance and user acceptance of such devices. The feedback provided is typically based on traditional eco-driving recommendations such as "accelerate gently" and "avoid braking". A more effective approach may be to provide drivers with predictive, feedforward information based on their upcoming needs (Jamson et al., 2015; McIlroy et al., 2017). Predictive systems may incorporate constraints relating to other driving objectives such as safety and collision avoidance (Dehkordi et al., 2019). Previous research has demonstrated that a combination of feedforward advice and feedback is more effective than either feedforward or feedback alone (Vaezipour et al., 2018). This chapter evaluates an assistance system that provides feedforward information based on a model of vehicle fuel consumption, prediction of the position of the preceding vehicle given its present velocity, and upcoming road geometry (Fleming et al., 2018). Novelly, this system incorporates a model of typical driver preferences on aspects such as following distances and cornering speeds to "coach" the driver into more fuel-efficient behaviour as a modification of their natural driving style (Fleming et al., 2020). Using a repeated-measures study in a fixed-base driving simulator, its efficacy is compared to traditional eco-driving advice in terms of fuel economy, effects on driver speed, acceleration and braking, and measures of cognitive workload. This is achieved by evaluating these metrics in three conditions: everyday driving (to act as a control condition), unassisted eco-driving, and assisted eco-driving.

HYPOTHESES

Several hypotheses were generated for the current work:

1. Both assisted and unassisted eco-driving will record reduced fuel usage compared to everyday driving, consistent with alterations in driving style.
2. For both assisted and unassisted eco-driving, participants' driving styles will be altered similarly to common eco-driving recommendations such as gentle acceleration and avoidance of braking.
3. Use of the assisted eco-driving will not be more effortful than everyday driving, therefore will not induce workload greater than everyday driving. In contrast, unassisted eco-driving will be associated with an increase in workload.

METHOD

Design

This study used a 3 (Condition) × 3 (Location) × 2 (Gender) mixed factorial design. The independent variables were Condition (Control/Unassisted Eco-driving/Assisted Eco-driving), Location (Urban/Rural/Motorway) [Within], and Gender (Male/Female) [Between]. The dependent variables were overall fuel use, as measured in kilograms (kg), speed, acceleration, braking dynamics, and workload.

Participants

36 participants (18 Males, 18 Females), aged 19–71 years ($M = 28.92$, $SD = 12.82$) completed this study. All participants held a current full driving license (UK or International) and all participants had normal or corrected normal vision. Participants were required to complete the Ishihara Colour Blindness Test prior to completing this study to ensure eligibility. Ethical approval for this study was given by the University of Southampton Research Governance Team (ERGO Number 30746).

Equipment and Driving Scenario

The primary piece of equipment used was the Southampton University Driving Simulator (SUDS). SUDS consists of a fixed base right-hand drive Land Rover Discovery with an automatic transmission. Three screens were positioned in front of

FIGURE 8.1 The University of Southampton driving simulator.

the vehicle to capture the forward projection of the road scene with 135-degree field of view. A single screen behind the vehicle showed a projected rear-view image that could be seen in the vehicle's rear-view mirror. In addition, the vehicles' side mirrors presented projections of the road environment through LCD screens. Figure 8.1 presents a view of the SUDS laboratory. STISIM M500W wide-field-of-view driving simulation software was used as the simulation program. Throughout the drive, participants' interactions, including speed, acceleration, headway, throttle position, and brake position were automatically logged by the simulator.

The simulated journey was modelled on a real-world route surrounding the University of Southampton, Hampshire, UK and included realistic road curvature, traffic density for a mid-afternoon journey, and localised landmarks. Development of this route is documented in Allison et al. (2017). This route included the different road types of urban, rural, and motorway sections of equal length. The route was 13 miles long and a drive lasted approximately 25 minutes. Urban areas of the route were interspersed at the beginning and end of the journey, whilst the motorway and rural sections appeared sequentially, separated by large roundabouts. Participants encountered six roundabouts as part of the route, which acted as junction decision points, whereby participants were required to significantly slow or come to a halt. There were two sets of pedestrian controlled traffic lights within the route, however these were set to green within the simulation. A map of the route participants was asked to drive is presented in Figure 8.2. Traffic levels encountered were based on early afternoon levels with no congestion, with a stable level of traffic flow in the urban situation (approx. 10–20 vehicles per mile) and free-flowing traffic in the motorway and rural situations (<10 vehicles per lane per mile).

As the simulator is a static vehicle, a digital speedometer was required to inform participants of their current speed. This was included in a visual interface that was developed in C# using Windows Presentation Foundation as a graphical library to display speed and RPM to the driver, based on real-time data from the STISIM

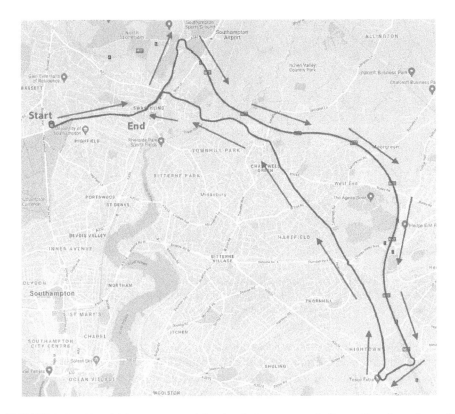

FIGURE 8.2 The driving route used in simulation, arrows indicate direction of travel. © 2021 Google.

simulation program. The resulting application was executed on a Microsoft Surface Pro tablet, which was placed behind the steering wheel of the car to replace the physical instrument cluster. The visual design of the interface is shown in Figure 8.3 and was comprised of a speed and RPM display.

To provide a speed recommendation to the driver, the instrument cluster display was augmented with a green and yellow "eco-band" to recommend a near-optimal speed range to the driver. The green region provided the participant with a range of recommended speeds for fuel-efficient driving based on the current road environment and situation, for example preceding cars and approaching corners. A further yellow region allowed for some sub-optimal speed selection. A similar "eco-speedometer" design was rated highly in perceived usefulness and user acceptance in previous research, however previous research lacked empirical testing of effectiveness (Meschtscherjakov et al., 2009). Further details regarding the development of this display are recorded in Allison and Stanton (2020).

When in use, the interface updates in real-time with the green eco-band stretching from zero speed to the current speed recommendation. This recommendation was generated by solving, in real-time, a numerical model predictive control optimisation problem that chose the throttle and braking action of the vehicle over the next

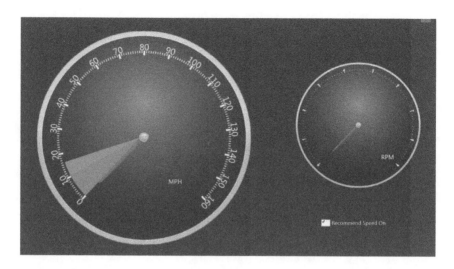

FIGURE 8.3 The speedometer used in the assisted eco-driving trial to display the recommended speed range. In the control and unassisted eco-driving conditions, the same display was used, but without the speed recommendation band.

30 seconds to minimise fuel consumption, accounting for typical driver preferences on speed and following distances. To predict the position of the vehicle ahead of the driver over this 30 second horizon, the simplification is used that the leader vehicle will maintain its current speed. Road geometry was also considered, with curves driven below a certain speed for driver comfort and safety. The current speed recommendation is taken as the vehicle speed at 10s into this 30 second prediction horizon and was updated continuously as the optimisation problem is resolved several times per second. The speed recommendation on the visual display was linearly interpolated between updates. The result is that the assistance system smoothly "coaches" the driver into following an efficient speed profile, encouraging such behaviours as coasting down to gradually reduce speed before sharp corners or behind stopped vehicles at intersections. As this optimisation incorporated a model of vehicle energy losses due to braking, aerodynamics, and rolling resistance, acceleration behaviour is recommended that leads to greater fuel efficiency. A detailed technical description may be found in Fleming et al. (2018).

Procedure

Upon entering the laboratory, participants were presented with an information sheet describing the aims of the research and outlining their ethical rights and a consent form. The experimenter verbally outlined the study to the participants and verbally restated their ethical rights. Participants were required to provide informed consent, and evidence of eligibility before the study could begin.

Prior to the start of the study, the driving simulator was adjusted to suit the individual participant's needs. Participants completed two practice drives within the simulator prior to the start of data collection. During these drives, participants were informed that the primary aim was to become accustomed to the simulator, the speedometer display, and to acclimatise to the speed cues which were offered by the environment, for example the engine noise and optic flow pattern. Both practice drives used the same urban route lasting approximately 10 minutes. For the second practice drive, participants had access to the eco-driving assistance system. Participants were told that the recommended speed would reduce fuel usage if followed, but were not given details of how the recommendation was made, nor that it was a 10 second preview. Data from the practice trials were not analysed. Following completion of the practice drives, participants proceeded to the main study.

Three different driving conditions were assessed as part of the study: a control condition in which drivers were asked to use their normal driving style, an unassisted eco-driving condition, whereby participants were presented with a series of eco-driving tips, including gentle acceleration, gentle braking, and anticipation of traffic flow, prior the condition but were not provided guidance during the drive (unassisted) and assisted eco-driving condition (assisted) in which the speed recommendation system described previously was used. To overcome potential order effects, the study was fully counterbalanced. Following completion of each simulated drive, participants completed the NASA-Task Load Index (NASA-TLX) (Hart & Staveland, 1988). Participants were encouraged to take a break following each drive, during which time the simulator reset.

Once participants had completed the three test trials and accompanying NASA-TLX, they were debriefed. The experimenter explained the aim of the study and presented participants with a debriefing form and compensation for their time. Participants were free to ask the experimenter questions about the study before leaving.

Measures and Analysis

STISIM simulation recorded numerous metrics based on participants' actions within the simulation. Participants' velocity, acceleration, throttle input, and braking input were sampled at a rate of 10Hz. This provided time-series data that could then be post-processed into summary statistics to evaluate a participant's driving style. Denoting the instantaneous values of velocity, acceleration, and braking respectively as v, a, and b, these summary statistics included the average (mean) speed:

$$\bar{v} = \frac{1}{T} \int_0^T v \, dt$$

where T is the total journey time. The integral appearing in this expression, and other integrals appearing in this section, were approximated using the trapezoidal rule during processing of the time-series data. To give an indication of whether the driver was travelling at a steady speed or accelerating and decelerating often, the root-mean-square (RMS) deviation of the speed from the mean was also calculated as:

$$v_{RMS} = \sqrt{\frac{1}{T}\int (v-\bar{v})^2 \, dt}$$

To evaluate if participants changed their acceleration behaviour, the time T_a during which the vehicle was accelerating above a small threshold value (0.5 m/s^2) was also considered. The mean acceleration during that time was then calculated as:

$$\bar{a} = \frac{1}{T_a} \int_{T_a} a \, dt$$

where the T_a under the integral sign indicates that the average over only those times for which the vehicle was accelerating was taken. As both mean acceleration and time spent accelerating are included as metrics, it is possible to distinguish between hard accelerations over short periods of time, and lower accelerations over longer times. Similarly, the time T_b during which the vehicle was braking was recorded, and the mean deceleration under braking was calculated as:

$$\bar{b} = \frac{1}{T_b} \int_{T_b} b \, dt$$

where again, the idea is that if drivers are braking more strongly or for longer in any condition, this should be clear from the calculated statistics.

Fuel usage was evaluated in post-processing using Simulink, using a detailed simulation model of the vehicle powertrain to estimate the fuel usage for each trial. This was carried out using the common "quasi-static" approach which generates a power demand profile to follow the vehicle speed data obtained from the simulation (Guzzella & Sciarretta, 2007). The model included a realistic engine map of fuel usage based on engine torque and RPM generated from one-dimensional fluid dynamics simulations in Ricardo Wave, and additionally incorporated efficiency losses in the transmission, a typical gear-switching logic for an automatic gearbox, aerodynamic losses due to vehicle drag, and losses due to rolling resistance. The powertrain model used in the optimisation to find the recommended speed for the eco-driving assistance system was a simplified version of this simulation model that ignored powertrain energy losses.

In addition to metrics gained from the simulator relating to driving style and fuel use, participants completed the NASA-TLX (NASA, 1986; Hart & Staveland, 1988) following each main study drive. This is a multivariate index to measure participants' subjectively perceived workload (Hart & Staveland, 1988) and is one of the most commonly used techniques to assess workload in practice (Stanton et al., 2013). It is comprised of six individual indices, designed to measure the "Mental", "Physical", "Temporal Demand", "Performance", "Effort", and "Frustration" components of overall cognitive load. NASA-TLX has been used across a variety of domains, including previous work exploring eco-driving (Jamson et al., 2015), driver distraction (Horberry et al., 2006), and aviation (Stanton et al., 2017; Stanton et al., 2018).

RESULTS

An alpha level of .05 was set for all statistical comparisons. Kolmogerov-Smirnov tests were used to verify the assumption of normality, then a series of 3 Condition (Control/Unassisted Eco/Assisted Eco) × 3 Location (Urban/Rural/Motorway) × 2 Gender (Male/Female) mixed analysis of variance (ANOVAs) were used. The Greenhouse-Geisser correction was used when data violated the assumption of sphericity. Where appropriate, post hoc paired-sample t-tests were calculated, with the Bonferroni correction for multiple comparisons being implemented, so that the unadjusted p-values were multiplied by the number of pairwise comparisons made in each case (9) to give the adjusted p-values reported.

As driving style measures can be evaluated for each participant and calculated separately for each condition and road type (location), this makes it possible to evaluate differences in driving styles in each case. This is followed by an analysis of workload in each condition from the NASA-TLX.

ANOVAs

To aid readability of the results, for the analysis of all considered driving metrics, main effects and interactions from the completed ANOVAs are presented in Table 8.1. Of clear interest within this table is the effect of condition.

TABLE 8.1
ANOVA for Condition, Location, and Gender, Main Effects and Interactions

DV	Effect	df	F	P	η_p^2
Fuel Usage	Condition	1.67, 56.86	29.8	<0.001***	0.47
	Location	1.57, 53.20	204	<0.001***	0.86
	Gender	1, 34	1.23	0.252	–
	Condition × Location	4, 136	6.69	<0.001***	0.16
Average Speed	Condition	2, 68	47	<0.001***	0.58
	Location	1.32, 44.95	27.2	<0.001***	0.99
	Gender	1, 34	7.98	0.008*	0.19
	Condition × Location	2.33, 79.35	18	<0.001***	0.35
RMS Speed Deviation	Condition	1.71, 58.26	11.8	<0.001***	0.26
	Location	1.52, 51.65	41.1	<0.001***	0.55
	Gender	1, 34	1.09	0.303	–
	Condition × Location	2.96, 100.79	4.67	0.004**	0.12
Mean Acceleration	Condition	1.49, 50.69	34.5	<0.001***	0.5
	Location	1.45, 49.30	65.5	<0.001***	0.66
	Gender	1, 34	1.03	0.785	–
	Condition × Location	2.57, 87.62	16.9	<0.001***	0.33
Acceleration Time	Condition	1.57, 53.29	8.49	<0.001***	0.46
	Location	1.67, 56.71	33.1	<0.001***	0.49
	Gender	1, 34	<1.00	0.938	–
	Condition × Location	4, 136	4.53	0.002**	0.12

* $p < .05$, ** $p < .005$, *** $p < .001$

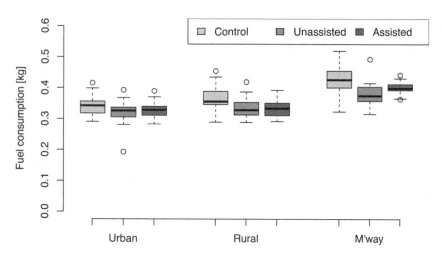

FIGURE 8.4 Participants mean fuel usage for the different locations, by condition.

Fuel Usage

Figure 8.4 presents participants' fuel usage by location for each of the test trials, and Table 8.2 presents results from the post hoc analysis. Results demonstrated that the unassisted eco-drive was associated with the lowest mean fuel usage and the control drive was associated with the highest mean fuel usage. It should be noted that there was only a significant difference in fuel usage between the unassisted and assisted eco-drives for the motorway section of the route.

Average Speed

Figure 8.5 (female (8.5a) and male (8.5b)) present participants' average speed, by location, for each of the test trials. These figures confirm that average speed was

TABLE 8.2
Post hoc Tests, Condition × Location Interaction for Fuel Usage

	Urban			Rural			Motorway		
	t(35)	p(adj)	MD	t(35)	p(adj)	MD	t(35)	p(adj)	MD
Unassisted - Control	−3.05	0.039*	−0.019	−4.82	<0.001***	−0.032	−7.52	<0.001***	−0.048
Assisted - Control	−2.83	0.070	−0.013	−4.89	<0.001***	−0.03	−4.49	<0.001***	−0.026
Assisted - Unassisted	1.28	1.000	0.006	0.294	1.000	0.001	4.25	0.001**	0.023

* $p < .05$, ** $p < .005$, *** $p < .001$

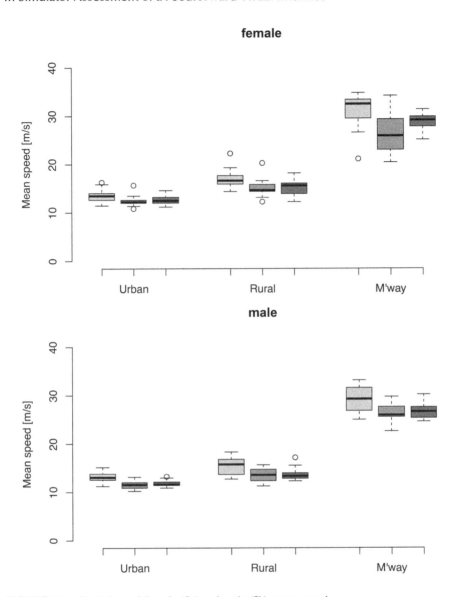

FIGURE 8.5 Participants' female (5a) and male (5b) mean speed.

highest during the motorway section of the route. The key difference between conditions was that participants travelled more slowly during the motorway section for the unassisted eco-driving condition. This is especially apparent when considering female participants, leading to the effect of gender identified within Table 8.1. When considering the post hoc analysis, Table 8.3 demonstrates that significant differences were seen between both unassisted and assisted eco-driving when compared to control. The only significant difference between unassisted and assisted eco-driving

TABLE 8.3
Post hoc Tests, Condition × Location Interaction for Mean Speed.

	Urban			Rural			Motorway		
	t(35)	p(adj)	MD	t(35)	p(adj)	MD	t(35)	p(adj)	MD
Unassisted - Control	−7.26	<0.001***	−1.38	−6.46	<0.001***	−1.88	−7.94	<0.001***	−4.09
Assisted - Control	−6.87	<0.001***	−1.17	−6.66	<0.001***	−1.79	−6.11	<0.001***	−2.53
Assisted - Unassisted	1.85	0.654	0.21	0.388	1.000	0.09	3.03	0.041*	1.55

* $p < .05$, ** $p < .005$, *** $p < .001$

occurred on the motorway, whereby unassisted eco-driving was associated with a lower mean speed.

SPEED RMS DEVIATION

Figure 8.6 presents participants' speed RMS deviation, by location, for each of the test trials. Whilst speed RMS deviation was typically highest during the rural section of the route, indicating greater changes in acceleration and deceleration, this is invariably linked to the route used and the nature of the rural roads within the simulation. It was seen that the control condition was typically associated with higher RMS speed deviation, significantly higher than both assisted and unassisted eco-driving in the rural section. When considering the effect of condition and location in more detail (Table 8.4), it appears that participants were able to keep more consistent speeds in the assisted eco-driving condition within the urban section of the drive than during the unassisted eco-drive.

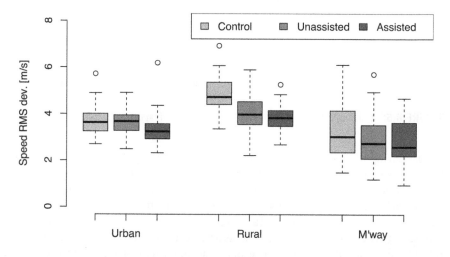

FIGURE 8.6 Participants' speed RMS deviation for the different locations, by condition.

TABLE 8.4
Post hoc Tests, Condition × Location Interaction for Speed RMS Deviation

	Urban			Rural			Motorway		
	t(35)	p(adj)	MD	t(35)	p(adj)	MD	t(35)	p(adj)	MD
Unassisted - Control	0.17	1.000	0.019	−4.52	<0.001***	−0.71	−2.02	0.459	−0.492
Assisted - Control	−2.55	0.137	−0.339	−6.79	<0.001***	−0.984	−2.04	0.444	−0.495
Assisted - Unassisted	−3.62	0.008*	−0.357	−2.04	0.444	−0.275	−0.01	1.000	−0.003

* $p < .05$, ** $p < .005$, *** $p < .001$

MEAN ACCELERATION

Figure 8.7 presents participants' mean acceleration, by location, for each of the test trials. This figure shows that greatest mean acceleration was seen in the control condition and lowest mean acceleration was recorded in the unassisted eco-driving condition, most apparently during the motorway section of the route. This difference was confirmed within the post hoc analysis (Table 8.5), which demonstrated

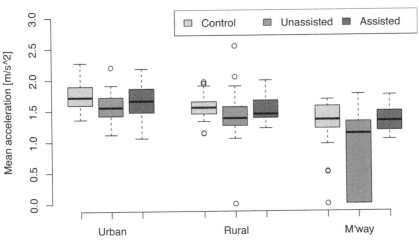

FIGURE 8.7 Participants' mean acceleration.

TABLE 8.5
Post hoc Tests, Condition × Location Interaction for Mean Acceleration

	Urban			Rural			Motorway		
	t(35)	p(adj)	MD	t(35)	p(adj)	MD	t(35)	p(adj)	MD
Unassisted - Control	−4.2	0.002**	−0.187	−2.52	0.147	−0.148	−4.57	<0.001***	−0.432
Assisted - Control	−1.96	0.518	−0.072	−0.86	1.000	−0.028	0.74	1.000	0.04
Assisted - Unassisted	2.69	0.099	0.115	1.9	0.588	0.12	4.72	<0.001***	0.472

* $p < .05$, ** $p < .005$, *** $p < .001$

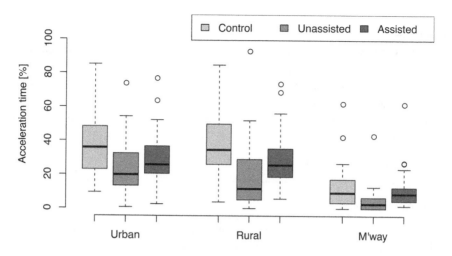

FIGURE 8.8 Participants' mean acceleration time in seconds.

significant differences between participants recorded mean acceleration for the control and unassisted eco-driving conditions for both the motorway and urban sections of the drive. Post hoc analysis also revealed a significant difference between mean acceleration for the assisted and unassisted eco-drivers during the motorway section.

ACCELERATION TIME

To tandardize the acceleration time metric across participants, this value was considered as a percentage of the journey time in each location. It is clear from Figure 8.8 that participants spent more time accelerating in the control condition for all locations. In contrast, considerably less acceleration time was seen in the unassisted eco-drive condition. Post hoc tests (Table 8.6) indicated significantly less acceleration in the unassisted eco-driving condition than either the control or assisted eco-drive.

MEAN BRAKING DECELERATION

Figure 8.9 presents mean braking deceleration for each of the conditions and locations explored within the current study. Braking deceleration was relatively

TABLE 8.6
Post hoc Tests, Condition × Location Interaction for Acceleration Time

	Urban			Rural			Motorway		
	$t(35)$	$p(adj)$	MD	$t(35)$	$p(adj)$	MD	$t(35)$	$p(adj)$	MD
Unassisted - Control	−5.04	<0.001***	−15.34	−5.53	<0.001***	−18.6	−5.33	<0.001***	−7.73
Assisted - Control	−3.4	0.016*	−8.97	−2.9	0.058	−9.15	−0.79	1.000	−1.24
Assisted - Unassisted	3.44	0.014*	6.38	3.16	0.029*	9.45	5.77	<0.001***	6.5

* $p < .05$, ** $p < .005$, *** $p < .001$

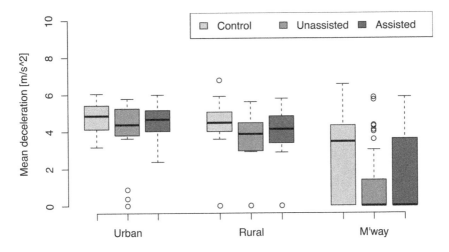

FIGURE 8.9 Participants' mean braking deceleration.

consistent in the different conditions and varied most noticeably with location, although was consistently lowest in the unassisted eco-drive trial. Post hoc tests (Table 8.7) reinforced the view that limited differences were observed between conditions and locations, with the exception that unassisted rural driving recorded significantly lower braking deceleration than the same location within the control condition.

Braking Time

As limited differences were observed relating to braking deceleration, overall braking time was also considered. It is clear from Figure 8.10 that the control condition was associated with the greatest mean braking time, with both eco-driving conditions (unassisted and assisted) being associated with less braking. Exploration of post hoc analysis (Table 8.8) demonstrated that both unassisted and assisted eco-drives were associated with significantly less braking time in both rural and urban sections of the route, compared to the control drives. The eco-driving conditions did not significantly differ from each other.

TABLE 8.7
Post hoc Tests, Condition × Location Interaction for Braking Deceleration

	Urban			Rural			Motorway		
	t(35)	p(adj)	MD	t(35)	p(adj)	MD	t(35)	p(adj)	MD
Unassisted - Control	−2.47	0.168	−0.55	−3.39	0.016*	−1.14	−2.84	0.068	−1.36
Assisted - Control	1.27	1.000	0.24	2.21	0.302	0.86	−1.76	0.777	−0.87
Assisted - Unassisted	−1.14	1.000	−0.31	−0.74	1.000	−0.28	−1.27	1.000	−0.49

* $p < .05$, ** $p < .005$, *** $p < .001$

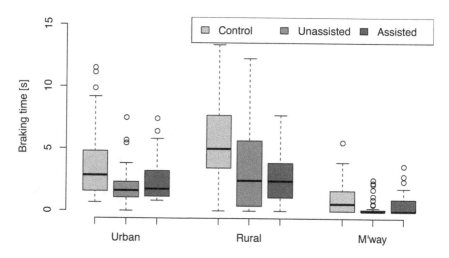

FIGURE 8.10 Participants' mean braking time in seconds.

WORKLOAD

In addition to objective data relating to participants' performance, subjective data was also collected relating to participants' overall level of workload following each drive, using the NASA-TLX. Each subscale of the NASA-TLX is presented separately and separately compared across the three conditions of the study using a series of 3 Condition (Control/Unassisted/Assisted) × 2 Gender (Male/Female) mixed ANOVAs. NASA-TLX scores recorded across the three conditions explored within the current study are presented in Figure 8.11.

For mental workload, a main effect of condition was identified, $F(2, 68) = 11.40$, $p < .05$, $\eta p^2 = .25$. Participants self-reported mental workload varied between the three trials. No effect of gender ($F < 1$) and no significant interaction between condition and gender ($F < 1$) was observed. Post hoc tests showed a significant difference between mental workload for control drive ($M = 8.83$, $SD = 5.39$) and the unassisted eco-driving drive ($M = 12.39$, $SD = 5.09$); $t(35) = -4.85$, $p = .001$. A significant difference was also identified in mental workload between the control drive and assisted

TABLE 8.8
Post hoc Tests, Condition × Location Interaction for Braking Time

	Urban			Rural			Motorway		
	t(35)	p(adj)	MD	t(35)	p(adj)	MD	t(35)	p(adj)	MD
Unassisted - Control	−3.61	0.009*	−1.75	−3.22	0.025*	−2.25	−2.67	0.104	−0.69
Assisted - Control	3.22	0.025*	1.41	6.17	<0.001***	2.96	2.24	0.286	0.51
Assisted - Unassisted	−1.28	1.000	−0.34	1.2	1.000	0.71	−1	1.000	−0.18

* $p < .05$, ** $p < .005$, *** $p < .001$

In Simulator Assessment of a Feedforward Visual Interface 185

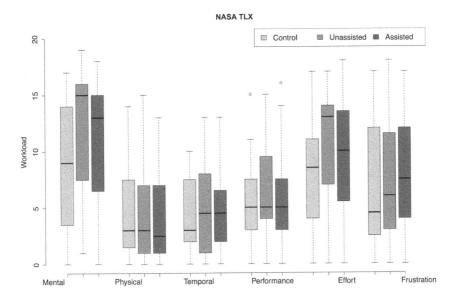

FIGURE 8.11 Participants' workload as measured by NASA-TLX.

eco-drive ($M = 11.47$, $SD = 4.92$), $t(35) = -3.38$, $p = .005$. There was no significant difference between the unassisted and assisted conditions; $t(35) = 1.13$, $p = .26$, ns.

Examination of participants effort ratings demonstrated a significant main effect of condition ($F(2, 68) = 7.42$, $p < .05$, $\eta p^2 = .18$), suggesting that condition impacted participants rating of required effort. No effect of gender ($F < 1$, ns) or interaction between condition and gender ($F < 1$, ns) was observed suggesting that gender did not influence effort ratings. Post hoc tests revealed a significant difference in participants ratings of effort for the control drive ($M = 8.22$, $SD = 4.01$) and the unassisted eco-drive ($M = 10.86$, $SD = 4.52$); $t(35) = -3.67$, $p = .001$, participants rated the required effort higher in the unassisted eco-drive than the control drive. However, there was no significant difference in effort rating between the control drive and assisted eco-drive ($M = 9.36$, $SD = 4.88$); $t(35) = -1.96$, $p = .174$, ns, or between the unassisted and assisted eco-drives; $t(35) = 2.05$, $p = .144$, ns.

All remaining subscales of the NASA-TLX failed to identify a main effect of condition (Physical ($F < 1$, ns), Temporal ($F < 1$, ns), Performance ($F < 1$, ns), Frustration ($F(1.71, 58.02) = 1.15$, ns).

DISCUSSION

Results suggest that participants' fuel usage was significantly reduced both when they were required to eco-drive without assistance and when provided with the speed recommendation system. Greatest fuel savings were observed within the unassisted eco-driving trial. Results suggest that the induced savings in both unassisted and assisted trials were due to a variety of factors, including travelling at a reduced speed and recording more consistent speeds throughout the drives.

Despite the fuel savings seen within the unassisted eco-driving trial, this was seen at the cost of greater mental workload, specifically higher self-reported effort. In contrast, whilst the assisted drive was associated with higher mental workload than the control drive, participants did not rate this interface as any more effortful than the control drive.

Prior to discussion of the differences between conditions, it is worth considering the effects of location that were identified within the current study. The finding that participants' fuel usage was greatest on the motorway, followed by the rural routes, followed by urban routes was as expected. This is because fuel usage is fundamentally linked to the speeds that participants would travel on these roads, which are signposted at 70 mph, 40 mph, and 30 mph respectively. Indeed, results of the current study demonstrate that location was a key determinant of participants' general driving behaviours. This reinforces the view that drivers are active in modifying their behaviours based on the environment they find themselves in, which vary based upon a variety of factors including legal speed limits, road curvature, and traffic conditions. Participants' ability to adapt their driving behaviour was also seen between the different conditions. Participants' fuel savings in both the assisted and unassisted eco-driving conditions reflected fundamental changes in driving style by the participants. To examine this in greater detail, participants' acceleration and braking dynamics were considered.

EFFECTS ON DRIVING STYLE

The results for mean acceleration confirm the view that the unassisted eco-drive was associated with significantly lower mean acceleration than the control drive. Interestingly other comparisons failed to reach required levels of significance. This suggests that participants used more gentle acceleration in the unassisted condition than the control condition, but not gentler than in the assisted condition. Of interest however was the limited acceleration participants used when on the motorway, suggesting that participants accelerated to a desired speed and then sought to maintain this speed, actions not possible on both the urban and rural sections of the route due to the presence of other road users, road geometry, and layout. The fact that the acceleration time was lower in the unassisted eco-driving condition also suggests that participants actively minimised their acceleration in this case.

It was seen that participants' acceleration profiles differed between conditions, indeed participants recorded greatest acceleration time within the control condition and least within the unassisted condition. As the acceleration profile also impacted fuel usage, it appears that the band was not as effective at influencing participants' acceleration as unassisted eco-driving was. Rather it appears that the assistive system was ineffective at moderating participants' acceleration, with participants rapidly accelerating to reach the maximum speed. This suggests further work is needed to investigate how to encourage participants to engage in gentle acceleration, specifically how the display can be refined to encourage participants to gently accelerate towards the recommended speed. The fact that participants were able to modify their behaviour to minimise fuel use reinforces that participants are able to rapidly and effectively eco-drive following minimal guidance, but require encouragement to initially engage in these behaviours.

EFFECTS ON FUEL CONSUMPTION

Participants' fuel usage was significantly reduced both when they were required to eco-drive without assistance and when they had access to the feedforward speed optimisation system. These savings were on average 11% and 8.5% of the fuel used in the control trial, respectively. The additional fuel savings recorded within the unassisted condition were associated with participants spending significantly less time accelerating within this trial compared to the other conditions, demonstrating that participants modified their driving behaviour beyond simply slowing down and maintaining consistent speeds.

Despite all the additional savings associated with both the assisted and unassisted eco-driving beyond slowing, it would be remiss to ignore that participants travelled at a lower speed during both the unassisted and assisted eco-driving conditions, particularly on the motorway. This causes lowering of losses from aerodynamic drag, which is typically a considerable contribution to overall fuel consumption at speeds of greater than 50 mph (Gillespie, 1992). The fuel savings seen within the assistive condition within the motorway of the route were not expected, as although the developed system is reactive to road events and curvature, the motorway is characterised by free-flowing traffic and limited curvature, and therefore would recommend a speed close to the speed limit of 70 mph. However, the system recommending that participants travelled at or below the speed limit acted to reinforce legal limits, limits that participants typically exceeded in the control condition, making this drive more legal, potentially safer, and offering reduced emissions as a by-product of the reduced maximum speed.

EFFECTS ON COGNITIVE WORKLOAD

Despite the fuel savings induced by unassisted eco-driving, it was seen that this was at the cost of greater mental workload and higher self-reported effort, supporting previous research (Pampel et al., 2015; Allison et al., 2020). In contrast, whilst assisted eco-driving was also associated with higher mental workload than the control drive, this was not statistically significant. This finding conflicts with previous research examining interface use and eco-driving which suggested that eco-driving with a visual interface induces greater workload than unassisted eco-driving (Jamson et al., 2015). Similarly, Lee et al. (2010) documented increased workload when participants were asked to use a visual interface designed to promote eco-driving. This suggests that the design of visual eco-driving assistance has a fundamental effect not only on its effectiveness in terms of fuel reduction but also on how hard participants must work to achieve this saving. Despite the overall positive results, no statistical difference was observed in effort ratings between the assisted and unassisted eco-driving, which may not have been the case if a greater sample size was used in the study. Future work is therefore needed to develop the visual interface to facilitate participants' eco-driving and explore participants' subsequent interactions in more detail.

LIMITATIONS

As a simulator-based study, there are several limitations within the present work that should be acknowledged. The first key limitation is that a model of fuel efficiency was used to measure fuel use, which may differ from real-world usage. Secondly, the study did not examine participants' qualitative opinions of the assistive system, which could provide useful feedback regarding whether participants liked such a representation of recommended speed and whether they would be willing to interact with the speed recommendation system outside of the simulator and controlled laboratory conditions. Consideration of the potential negative effects of the speed recommendation system is also needed prior to on-road testing, where effects such as distraction linked to the speed recommendation system could have considerable safety implications.

OPPORTUNITIES FOR FUTURE WORK

This study has highlighted several avenues for further research. Firstly, this study has revealed that despite the improvements offered by the assistance system in terms of fuel use, this could be improved further to more closely reflect the fuel use recorded within the eco-trial. This improvement could be achieved by providing drivers with an indication of when to release the accelerator to encourage greater coasting, reducing the need for braking, and guidance regarding the rate to accelerate to reach a recommended speed. Previous approaches to encouraging greater coasting behaviour have combined the use of haptic information through the accelerator with other stimuli to provide participants information on the best time to reduce acceleration (Jamson et al., 2013). It is therefore possible a system such as the one considered in the current study could be combined with alternative feedback modalities.

In addition to opportunities relating to refinements to the assistance system, there is the potential and need to test the feedforward speed optimisation system within the real road environment. Two hurdles must be addressed for this to be viable. The first is suitable integration with on-board vehicle sensors, such as radar and LIDAR systems, and the vehicle engine management system. The second obstacle is ensuring safe, accurate, and meaningful predictions from the system within a volatile road environment. Within the simulated conditions explored all traffic behaviour is pre-scripted and hence predictable, but in the real-world this is not the case. The increasing potential for future connected and autonomous vehicles (Banks et al., 2018) however provides ideal impetus for the need to develop and utilise feedforward technologies, as explored within the current work.

CONCLUSION

The desire to reduce emissions and increase the fuel economy associated with road vehicles is shared by governments, automotive manufacturers, and drivers. This study demonstrated success in the assessment of a novel feedforward speed optimisation display to coach a driver into fuel-efficient driving and provided evidence that such a device can be used to reduce fuel usage compared to everyday driving behaviours. Although the display was not as effective at reducing total fuel usage as when participants were specifically asked to eco-drive, it was found that participants' success at unassisted eco-driving was at the cost of increased self-reported effort suggesting that participants may struggle to maintain this behaviour outside of laboratory conditions. Further refinement of the visual interface is possible to facilitate its use.

REFERENCES

Allison, C. K., Fleming, J. M., Yan, X., Lot, R., & Stanton, N. A. (2020). Adjusting the need for speed: Assessment of a visual interface to reduce fuel use. *Ergonomics*, *64* (3), 315–329.

Allison, C. K., Parnell, K. J., Brown, J. W., & Stanton, N. A. (2017). Modeling the real world using STISIM drive® simulation software: A study contrasting high and low locality simulations. *International Conference on Applied Human Factors and Ergonomics*, (pp. 906–915).

Allison, C. K., & Stanton, N. A. (2020). Ideation using the "Design with intent" toolkit: A case study applying a design toolkit to support creativity in developing vehicle interfaces for fuel-efficient driving. *Applied Ergonomics*, *84*, 103026.

Banks, V. A., Stanton, N. A., Burnett, G., & Hermawati, S. (2018). Distributed cognition on the road: Using EAST to explore future road transportation systems. *Applied Ergonomics*, *68*, 258–266.

Barkenbus, J. N. (2010). Eco-driving: An overlooked climate change initiative. *Energy Policy*, *38*, 762–769.

Beusen, B., Broekx, S., Denys, T., Beckx, C., Degraeuwe, B., Gijsbers, M., & Panis, L. I. (2009). Using on-board logging devices to study the longer-term impact of an eco-driving course. *Transportation Research Part D: Transport and Environment*, *14*(7), 514–520.

Birol, F. (2011). *"CO_2 emissions from fuel combustion–highlights."* https://www.osti.gov/etdeweb/servlets/purl/21589332

Birrell, S. A., Young, M. S., & Weldon, A. M. (2013). Vibrotactile pedals: Provision of haptic feedback to support economical driving. *Ergonomics*, *56*, 282–292.

Boriboonsomsin, K., Vu, A., & Barth, M. (2010). *Eco-driving: Pilot evaluation of driving behavior changes among us drivers*. Berkeley, CA University of California Transportation Center.

Botkin, D. B., Saxe, H., Araujo, M. B., Betts, R., Bradshaw, R. H., Cedhagen, T., Chesson, P., Dawson, T. P., Etterson, J. R., Faith, D. P., & Ferrier, S. (2007). Forecasting the effects of global warming on biodiversity. *AIBS Bulletin*, *57*(3), 227–236.

Chan, C. C. (2007). The state of the art of electric, hybrid, and fuel cell vehicles. *Proceedings of the IEEE*, *95*, 704–718.

Dehkordi, S. G., Larue, G. S., Cholette, M. E., Rakotonirainy, A., & Rakha, H. A. (2019). Ecological and safe driving: A model predictive control approach considering spatial and temporal constraints. *Transportation Research Part D: Transport and Environment*, *67*, 208–222.

Delhomme, P., Cristea, M., & Paran, F. (2013). Self-reported frequency and perceived difficulty of adopting eco-friendly driving behavior according to gender, age, and environmental concern. *Transportation Research Part D: Transport and Environment*, *20*, 55–58.

Dogan, E., Bolderdijk, J. W., & Steg, L. (2014). Making small numbers count: Environmental and financial feedback in promoting eco-driving behaviours. *Journal of Consumer Policy*, *37*, 413–422.

Fischer, G., Shah, M., N. Tubiello, F., & Van Velhuizen, H. (2005). Socio-economic and climate change impacts on agriculture: An integrated assessment, 1990–2080. *Philosophical Transactions of the Royal Society B: Biological Sciences, 360*(1463), 2067–2083.

Fleming, J., Yan, X., & Lot, R. (2020). Incorporating driver preferences into eco-driving assistance systems using optimal control. *IEEE Transactions on Intelligent Transportation Systems. 22*(5), 2913–2922.

Fleming, J. M., Yan, X., Allison, C. K., Stanton, N. A., & Lot, R. (2018). Driver modeling and implementation of a fuel-saving ADAS. *IEEE Conference on Systems, Man and Cybernetics (SMC)*. (pp. 1233–1238). IEEE.

Froehlich, J., Dillahunt, T., Klasnja, P., Mankoff, J., Consolvo, S., & Harrison, B. (2009). UbiGreen: Investigating a mobile tool for tracking and supporting green transportation habits. *Proceedings of the SIGCHI Conference on Human Factors in Computing Systems*, (pp. 1043–1052).

Gillespie, T. D. (1992). *Vehicle Dynamics*. SAE International.

Guzzella, L., & Sciarretta, A. (2007). *Vehicle propulsion systems* (Vol. *1*). Berlin-Heidelberg, Germany: Springer-Verlag.

Hart, S. G., & Staveland, L. E. (1988). Development of NASA-TLX (task load index): Results of empirical and theoretical research. In: *Advances in psychology* (Vol. *52*, pp. 139–183). Elsevier, North-Holland.

Heyes, D., Daun, T. J., Zimmermann, A., & Lienkamp, M. (2015). The virtual driving coach-design and preliminary testing of a predictive eco-driving assistance system for heavy-duty vehicles. *European Transport Research Review, 7*(3), 25.

Horberry, T., Anderson, J., Regan, M. A., Triggs, T. J., & Brown, J. (2006). Driver distraction: The effects of concurrent in-vehicle tasks, road environment complexity and age on driving performance. *Accident Analysis & Prevention, 38*(1), 185–191.

IEA. (2017). *Key world energy statistics*. From https://www.iea.org/publications/freepublications/publication/KeyWorld2017.pdf.

Jamson, A. H., Hibberd, D. L., & Merat, N. (2013). The design of haptic gas pedal feedback to support eco-driving. In: *Proceedings of the seventh international driving symposium on human factors in driver assessment, training, and vehicle design* (pp. 264–270). University of Iowa, Iowa City, Iowa.

Jamson, S. L., Hibberd, D. L., & Jamson, A. H. (2015). Drivers' ability to learn eco-driving skills: Effects on fuel efficient and safe driving behaviour. *Transportation Research Part C: Emerging Technologies, 58*, 657–668.

Larue, G. S., Malik, H., Rakotonirainy, A., & Demmel, S. (2014). Fuel consumption and gas emissions of an automatic transmission vehicle following simple eco-driving instructions on urban roads. *IET Intelligent Transport Systems, 8*(7), 590–597.

Lee, H., Lee, W., & Lim, Y. K. (2010). The effect of eco-driving system towards sustainable driving behavior. In *CHI'10 Extended Abstracts on Human Factors in Computing Systems* (pp. 4255–4260).

Lorf, C., Martínez-Botas, R. F., Howey, D. A., Lytton, L., & Cussons, B. (2013). Comparative analysis of the energy consumption and CO_2 emissions of 40 electric, plug-in hybrid electric, hybrid electric and internal combustion engine vehicles. *Transportation Research Part D: Transport and Environment, 23*, 12–19.

Martin, E., Chan, N., & Shaheen, S. (2012). How public education on eco-driving can reduce both fuel use and greenhouse gas emissions. *Transportation Research Record: Journal of the Transportation Research Board, 2287*, 163–173.

McIlroy, R. C., Stanton, N. A., & Harvey, C. (2013). Getting drivers to do the right thing: A review of the potential for safely reducing energy consumption through design. *IET Intelligent Transport Systems, 8*(4), 388–397.

McIlroy, R. C., Stanton, N. A., Godwin, L., & Wood, A. P. (2017). Encouraging eco-driving with visual, auditory, and vibrotactile stimuli. *IEEE Transactions on Human-Machine Systems*, *47*(5), 661–672.

McMichael, A. J., Woodruff, R. E., & Hales, S. (2006). Climate change and human health: Present and future risks. *The Lancet*, *367*(9513), 859–869.

Mensing, F., Bideaux, E., Trigui, R., Ribet, J., & Jeanneret, B. (2014). Eco-driving: An economic or ecologic driving style? *Transportation Research Part C: Emerging Technologies*, *38*, 110–121.

Meschtscherjakov, A., Wilfinger, D., Scherndl, T., & Tscheligi, M. (2009). Acceptance of future persuasive in-car interfaces towards a more economic driving behaviour. *Proceedings of the 1st International Conference on Automotive User Interfaces and Interactive Vehicular Applications*, (pp. 81–88).

NASA-Task Load Index (TLX). (1986). *Computerized Version*. Moffett Field. CA: NASA-Ames Research Center, Aerospace Human Factors Research Division.

Pampel, S. M., Jamson, S. L., Hibberd, D. L., & Barnard, Y. (2015). How I reduce fuel consumption: An experimental study on mental models of eco-driving. *Transportation Research Part C: Emerging Technologies*, *58*, 669–680.

Piatkowski, D., Bronson, R., Marshall, W., & Krizek, K. J. (2014). Measuring the impacts of bike-to-work day events and identifying barriers to increased commuter cycling. *Journal of Urban Planning and Development*, *141*, 04014034.

Rose, G., & Marfurt, H. (2007). Travel behaviour change impacts of a major ride to work day event. *Transportation Research Part A: Policy and Practice*, *41*, 351–364.

Saboohi, Y., & Farzaneh, H. (2009). Model for developing an eco-driving strategy of a passenger vehicle based on the least fuel consumption. *Applied Energy*, *86*(10), 1925–1932.

Skeie, R. B., Fuglestvedt, J., Berntsen, T., Lund, M. T., Myhre, G., & Rypdal, K. (2009). Global temperature change from the transport sectors: Historical development and future scenarios. *Atmospheric Environment*, *43*, 6260–6270.

Stanton, N. A., Harvey, C., Plant, K. L., & Bolton, L. (2013). To twist, roll, stroke or poke? A study of input devices for menu navigation in the cockpit. *Ergonomics*, *56*(4), 590–611.

Stanton, N. A., Plant, K. L., Roberts, A. P., & Allison, C. K. (2017). Use of highways in the sky and a virtual pad for landing head up display symbology to enable improved helicopter pilot's situation awareness and workload in degraded visual conditions. *Ergonomics*, *62*(2), 255–267.

Stanton, N. A., Plant, K. L., Roberts, A. P., Allison, C. K., & Harvey, C. (2018). The virtual landing pad: Facilitating rotary-wing landing operations in degraded visual environments. *Cognition, Technology & Work*, *20*(2), 219–232.

Tulusan, J., Staake, T., & Fleisch, E. (2012, September). Providing eco-driving feedback to corporate car drivers: What impact does a smartphone application have on their fuel efficiency? In: *Proceedings of the 2012 ACM conference on ubiquitous computing* (pp. 212–215). ACM.

Turrentine, T. S., & Kurani, K. S. (2007). Car buyers and fuel economy? *Energy Policy*, *35*(2), 1213–1223.

Vaezipour, A., Rakotonirainy, A., Haworth, N., & Delhomme, P. (2018). A simulator evaluation of in-vehicle human machine interfaces for eco-safe driving. *Transportation Research Part A: Policy and Practice*, *118*, 696–713.

Vining, C. B. (2009). An inconvenient truth about thermoelectrics. *Nature materials*, *8*, 83.

Windecker, A., & Ruder, A. (2013). Fuel economy, cost, and greenhouse gas results for alternative fuel vehicles in 2011. *Transportation Research Part D: Transport and Environment*, *23*, 34–40.

World Economic Forum (2018, August). What are the 10 biggest global challenges? Retrieved from https://www.weforum.org/agenda/2016/01/what-are-the-10-biggest-global-challenges.

Yilmaz, M., & Krein, P. T. (2013). Review of battery charger topologies, charging power levels, and infrastructure for plug-in electric and hybrid vehicles. *IEEE Transactions on Power Electronics, 28*, 2151–2169.

Young, M. S., Birrell, S. A., & Stanton, N. A. (2009, July). Design for smart driving: A tale of two interfaces. In: *International conference on engineering psychology and cognitive ergonomics* (pp. 477–485). Berlin, Heidelberg: Springer.

9 Assisted versus Unassisted Eco-Driving for Electrified Powertrains

INTRODUCTION

Motivated by climate change and the drive for sustainability, recent years have seen a great increase in measures to limit energy consumption and improve the energy efficiency of personal transportation. For combustion engine-driven vehicles, greater energy efficiency lowers carbon dioxide (CO_2) emissions, while for plug-in electric vehicles it can reduce charging costs and increase range. Eco-driving, the modification of driving behaviour to save fuel and energy, is an efficacious method to reduce vehicle energy usage (Barkenbus, 2010) that nonetheless can have limited effectiveness in practice due to difficulty in getting drivers to maintain energy-efficient behaviour long-term (Beusen et al., 2009). One suggestion to improve the uptake and retainment of eco-driving behaviour is to provide real-time feedback on a driver's energy efficiency through a visual (Meschtscherjakov et al., 2009) or auditory or haptic (McIlroy et al., 2016) human-machine interface (HMI). This real-time feedback may be augmented with predictive or feedforward information on the behaviour of traffic and traffic signals to improve energy saving beyond what can be achieved by the driver alone (Kamal et al., 2010). In addition, such systems show great promise for use in future autonomous vehicles (Wu et al., 2011), where the HMI may be neglected and commands sent directly to the autonomous driving software.

A possible complication of future eco-driving efforts is a trend towards electrification of vehicle powertrains (Tate et al., 2008), driven itself by the pursuit of greater energy efficiency. Typical eco-driving recommendations such as gentle acceleration, late upshifting of gears, and avoidance of braking (Barkenbus, 2010) were developed for combustion engine-driven vehicles. Electric powertrains can regenerate energy when braking and typically use an electric motor and a fixed reduction gear in place of a combustion engine and gearbox, leading to different optimal operating points from the perspective of energy efficiency. Nonetheless, the effect of driving style on the energy efficiency of electric vehicles has been studied, and eco-driving behaviours have shown to be effective for them (Knowles et al., 2012). These eco-driving behaviours may need to be adapted from those that are effective for conventional vehicles, supporting the use of either specific driver training or assistance systems (Neumann et al., 2015). Hybrid vehicles have also been considered in the literature, with specific recommendations made for the design of in-vehicle interfaces to promote eco-driving (Franke et al., 2016). From a technical standpoint, eco-driving

DOI: 10.1201/9781003081173-9

optimisations have been developed for electric vehicles (Dib et al., 2014), and for hybrid vehicles (Hu et al., 2016), which may also use vehicle-infrastructure communication to provide recommendations.

In this work, we evaluate the efficacy of different powertrains of a recently developed eco-driving assistance system that incorporates a predictive model of vehicle energy losses and, novelly, driver preferences in following and cornering situations (Fleming et al., 2018). This is carried out by calculating energy usage for velocity profiles obtained from human participants in a driving simulator at the University of Southampton, UK (Yan et al., 2020). Novelly, this assistance system provides speed recommendations based on a multi-objective optimisation of minimising energy loss and respecting typical driver preferences on vehicle speed, following distance, and cornering speed. This allows it to gently "coach" drivers into natural, yet more energy-efficient behaviour. Fuel and/or electrical energy usage may be calculated after simulation using models of conventional, hybrid electric, and electric powertrains. The objective of this analysis is to determine if the developed driver assistance system is effective across different powertrains, comparing it with normal driving and unassisted eco-driving in each case. As there are known gender differences in driving behaviour, especially regarding speed choice (Cestac et al., 2011), gender was also considered as a factor in our statistical study, although it did not correlate to any differences in energy consumption. We do not include an in-depth description of the assistance system and optimisation itself or report measures of driving style and cognitive workload. Instead, we direct readers to the authors' other works (Fleming et al., 2018; Allison et al., 2020; Yan et al., 2020) to cover these aspects.

POWERTRAIN MODELS

To evaluate energy consumption, we used quasi-static models of the three powertrains in Figure 9.1. These models were developed in Simulink and validated as part of a wider research project (Yan et al., 2017). For each participant, the velocity of the modelled vehicle was controlled to track the velocity profile of the participants in the driving simulator study using a proportional-integral-derivative (PID) feedback controller tuned to apply force to give less than 0.01 m/s mean-square error in velocity with no overshoot. Components of the three powertrains were sized so that each had a total power output of 150 kW, to match the characteristics of the driving simulation. All three models also considered equal driving energy losses due to rolling resistance and aerodynamic drag. Hence, in each case the overall driving force F on the wheels was:

$$F = ma_x + \frac{1}{2}\rho A_f C_d v^2 + C_{rr} mg \tag{9.1}$$

where m is the vehicle mass, ρ the air density, A_f the frontal area, C_d the coefficient of drag, and C_{rr} the coefficient of rolling resistance. Once the force required to follow the velocity profile was found, maps of fuel and electrical energy usage were evaluated using linear interpolation to find instantaneous fuel/energy consumption, which was integrated over the entire journey. Values of the parameters used in the study are given in Table 9.1.

Assisted versus Unassisted Eco-Driving for Electrified Powertrains

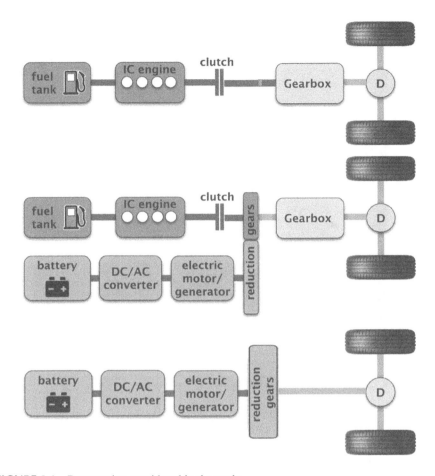

FIGURE 9.1 Powertrains considered in the study.

CONVENTIONAL POWERTRAIN

For the conventional internal combustion engine (ICE) powertrain (Figure 9.1a), the wheels are driven by a 150 kW gasoline engine via the gearbox. Gear changes are automatic and carried out according to a typical two-parameter gear shifting

TABLE 9.1
Vehicle Parameters

Parameter	Value	Parameter	Value
M	1500 kg	ρ	1.225 kg/m^3
G	9.81 N/kg	C_{rr}	0.008
C_d	0.4	r_w	0.3 m
A_f	2 m^2		

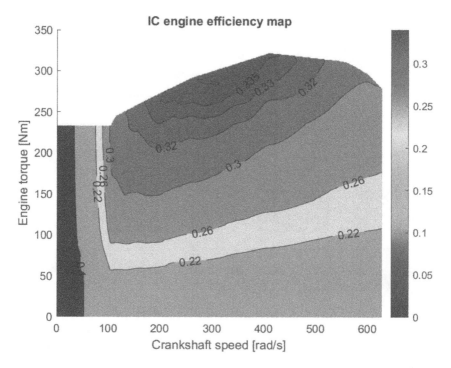

FIGURE 9.2 Efficiency map of the IC engine.

schedule. The overall driving force F is the sum of the force caused by the engine and the hydraulic brakes. Fuel consumption of the ICE is modelled as a lookup table with two inputs: crankshaft speed and engine torque, which are related to vehicle speed v and driving force F via the transmission. The efficiency map used for the ICE was obtained from 1d fluid dynamics simulations in Ricardo Wave and is given in Figure 9.2. As is typical for a gasoline engine, peak efficiency occurs at high torque and moderate revolutions per minute (RPM), and efficiency decreases rapidly below 100 rad/s due to clutch slip.

Parallel Hybrid Powertrain

The second powertrain considered is the parallel hybrid electric vehicle (HEV) (Figure 9.1b). In addition to the ICE, this includes a 15 kW electric traction motor, which is coupled to the driving shaft via a fixed reduction gear.

Hence, the driving force F is provided by three sources: the engine, the traction motor, and the hydraulic brakes. The fuel consumption model of the ICE is identical to that used for the conventional powertrain, while the electrical traction system is modelled in a similar manner using the efficiency map (Figure 9.3). The performance of the HEV is dependent on the energy management strategy used to control the power split between these sources. In this chapter, the well-known equivalent consumption minimisation strategy (ECMS) (Paganelli et al., 2001) has been used, which is optimal if the battery charge is equal at the start and end of the drive. The equivalence factor

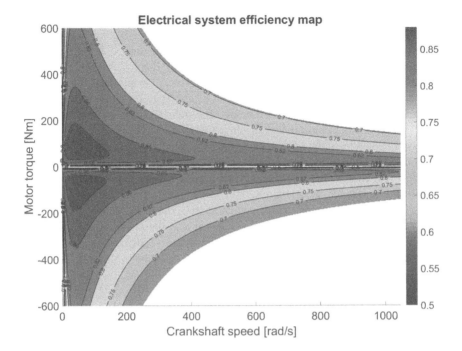

FIGURE 9.3 Efficiency map of the electric traction motor.

used in the ECMS algorithm was tuned to ensure this charge-sustaining behaviour, so that energy consumption may be evaluated considering fuel usage only.

Battery Electric Powertrain

The third powertrain is electric (Figure 9.1c) and uses a similar electrical traction system to that in the parallel HEV. However, the battery and motor have been resized from 15 kW to 150 kW to match the performance of the ICE vehicle. For this powertrain, the driving force F is provided by the motor and the hydraulic brakes. The motor may apply a negative force, allowing for regenerative braking, but this regenerative braking force is only applied on the front wheels so that a proportion of braking energy is recovered when the driver uses the brake pedal in the simulator, with manual braking assumed for braking forces greater than can be provided by the electric motor.

STUDY METHOD

Equipment

The study was carried out in the fixed-base driving simulator at the University of Southampton, which comprises: a right-hand drive Land Rover Discovery; three projector screens in front of the vehicle with a 135-degree field of view; a single screen behind the vehicle showing an image visible in the rear-view mirror, and LCD displays on each side-view mirror to simulate the view there.

A 21 km simulated route consisting of urban, rural, and motorway sections was developed and driven using STISIM M500W wide-field-of-view driving simulation software, with the journey modelled on real-world roads around Southampton, UK. More information on this route and its development may be found in (Allison et al., 2017).

An additional speedometer was required to show participants their speed. This was implemented as a graphical user interface (GUI) application executed on a Microsoft Surface Pro tablet, which was placed behind the steering wheel of the car to replace the physical instrument cluster and display speed and RPM. The eco-driving assistance system considered is an eco-speedometer design that augments this speedometer with a green and yellow band to recommend a speed range. This recommendation is updated in real-time to provide a speed that approximately minimises energy losses due to rolling resistance, aerodynamic drag, and braking. The efficiency of the drivetrain is not taken into account, and the calculation of the recommended speed considers the current speed and acceleration of the vehicle, the position and speed of any leading vehicle, and the curvature of approaching corners. Further details on the visual HMI and the predictive speed recommendation may be found in (Fleming et al., 2018).

STUDY DESIGN

36 participants (18 males, 18 females) were recruited, aged 19 to 71 years ($M = 28.92$, $SD = 12.82$).

To acclimatise to the simulator, they first completed two short practice drives, with the speed recommendation band used during the main study turned on for the second practice. The main study used a repeated-measures design with one within-subjects independent variable: condition (control/eco/assisted). Three conditions were considered:

- Control condition: The driver was instructed to drive as they normally would.
- Eco-driving condition: The driver was instructed to drive in order to save energy, after being shown a set of eco-driving tips, particularly to avoid braking, to accelerate smoothly and gently, to anticipate traffic flow, and observe speed limits.
- Assisted condition: The eco-driving assistance system was used to display a band of recommended speeds on the vehicle speedometer. Drivers were informed that the band recommended speeds for effective eco-driving, but were not explicitly instructed to follow the recommendation.

To reduce the influence of order effects on the results, the order that participants drove these conditions was fully counterbalanced with an equal number given each of the six possible permutations.

STATISTICAL ANALYSIS

The significance level of 0.05 was chosen for all comparisons in the study. For each powertrain, a repeated-measures analysis of variance (ANOVA) was used to analyse the dependence of fuel/electric energy consumption on test conditions. Tests of normality and sphericity were performed to check the assumptions of the ANOVA analysis. Where significant effects were observed, post hoc paired-sample t-tests were used to compare groups, applying a Bonferroni correction to account for multiple comparisons. Although the effect of gender was considered in the analysis, it did not correspond to significant differences in any case, and so we do not discuss it further.

RESULTS

Descriptive statistics and ANOVA results are given in Tables 9.2 and 9.3 respectively. Kolmogorov-Smirnov tests indicated that residuals were not significantly different from a normal distribution of mean zero ($p = 0.52$, $p = 0.36$, and $p = 0.56$ for the conventional, hybrid, and electric cases respectively). Mauchly's test indicated that the sphericity assumption of the ANOVA may be violated ($p = 0.026$, $p = 0.055$, and $p = 0.001$), hence Greenhouse-Geisser corrections were applied. A significant main effect of test condition was observed in all cases. Accordingly, we performed post hoc t-tests between different test conditions for each powertrain. The results of these post hocs may be seen in Table 9.4. Differences were significant in all cases except between the eco-driving and assisted condition for the hybrid powertrain.

Boxplots of the fuel and electrical energy consumption of the three powertrains in the study are shown in Figures 9.4–9.6, noting that the hybrid is operated in a charge-sustaining manner and its fuel usage represents the total energy use. Eco-driving led to the lowest consumption in all three cases. The benefit ranged from a reduction of 8.83% for the conventional powertrain, to 24.2% in the electric powertrain, and the hybrid intermediate between these with a 17.5% reduction. Use of the driver assistance system also showed greater benefits for the electrified

TABLE 9.2
Descriptive Statistics

	Control M	Control SD	Eco M	Eco SD	Assisted M	Assisted SD
Conv. [l/100 km]	7.02	0.53	6.40	0.43	6.59	0.33
Hyb. [l/100 km]	5.93	0.81	4.89	0.71	4.99	0.36
Elec. [kWh/100 km]	12.04	2.38	9.13	1.49	10.04	1.28

TABLE 9.3
ANOVA Results for Test Condition

Powertrain	F	P	η_p^2
Conventional	29.8	<0.001*	0.467
Hybrid	44.5	<0.001*	0.567
Electric	41.9	<0.001*	0.552

TABLE 9.4
Post Hoc t-Tests of Fuel/Energy Usage

Condition Pair	Mean Diff.	T	P
Conventional			
Control – Eco-driving	0.62 l/100 km	6.32	<0.001*
Control – Assisted	0.43 l/100 km	5.64	<0.001*
Assisted – Eco-driving	0.19 l/100 km	2.78	=0.009*
Hybrid			
Control – Eco-driving	1.05 l/100 km	7.34	<0.001*
Control – Assisted	0.95 l/100 km	8.24	<0.001*
Assisted – Eco-driving	0.10 l/100 km	0.96	=0.343
Electric			
Control – Eco-driving	2.90 kWh/100 km	7.38	<0.001*
Control – Assisted	1.99 kWh/100 km	6.19	<0.001*
Assisted – Eco-driving	0.91 kWh/100 km	3.97	<0.001*

powertrains. For the conventional powertrain, there was a 6.13% reduction in energy usage, but the hybrid and electric powertrains showed 15.9% and 16.6% reductions respectively.

Although it was not the objective of the study, we also note that the fuel consumption of the parallel hybrid drivetrain was substantially (15.5%–24.3%) lower than that of the conventional one for all trials, confirming that hybridisation still leads to benefits when eco-driving is used.

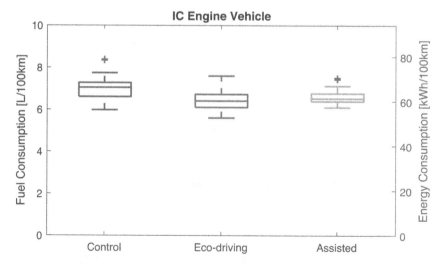

FIGURE 9.4 Fuel usage of conventional powertrain.

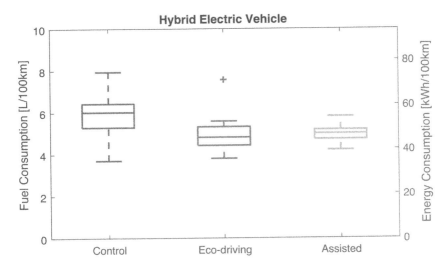

FIGURE 9.5 Fuel usage of hybrid powertrain.

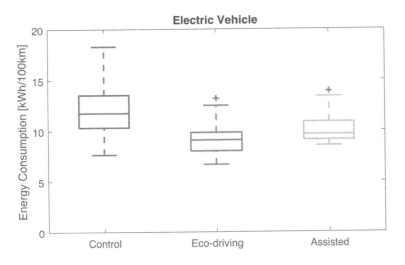

FIGURE 9.6 Energy usage of electric powertrain.

DISCUSSION

In line with existing publications discussing eco-driving for conventional (Barkenbus, 2010) and electric vehicles (Knowles et al., 2012), the results confirm that eco-driving advice to avoid braking, to accelerate smoothly and gently, to anticipate traffic flow, and to observe speed limits is useful to reduce energy consumption. As the same time-series data for velocity was used to evaluate each model, this also implies that generic eco-driving advice is efficacious to reduce drivers energy consumption in hybrid and battery electric vehicles, which may conflict with suggestions that eco-driving training should be modified for electrified powertrains (Neumann et al., 2015). Indeed, proportionally greater reductions are observed for the two electrified powertrains than for the conventional one.

The results also add to the existing literature that supports the use of eco-driving assistance using visual feedback for conventional (Meschtscherjakov et al., 2009; Kamal et al., 2010), hybrid (Franke et al., 2016), and electric vehicles (Neumann et al., 2015). The present work extends this literature by demonstrating that this is also the case for the assistance systems using predictive models of lead vehicle motion and driver preferences (Fleming et al., 2018), as considered here. However, the observation that the assistance system may be outperformed by drivers who actively eco-drive, particularly in the electric case, reinforces suggestions that assistance systems may perform better if they are designed specifically for electric vehicles (Neumann et al., 2015).

LIMITATIONS

As the study was limited to evaluating velocity data from each participant using quasi-static models, simulated vehicle dynamics were the same for all powertrains. Hence, although overall power and acceleration were matched to the simulated vehicle in each case, the present study cannot account for differences in driving style resulting from characteristics of the powertrains. As the study was carried out in a fixed-base simulator, driving style may also differ from real-world driving, for example due to the lack of lateral acceleration when cornering. We have not considered learning effects or changes over time, which is often an important aspect for eco-driving training and assistance systems. It is also possible that features of the simulator itself may have affected participants' adopted driving styles, adapting their style after seeing the Land Rover Discovery, for example.

An automatic transmission was assumed and the study did not incorporate the effect of drivers' gear-changing strategies. The driver assistance system used provided only a speed recommendation via a visual HMI and so the results may not generalise to other kinds of assistance systems. Finally, the hybrid powertrain considered was a pre-transmission parallel hybrid, and so the results may not generalise to series or series-parallel powertrains.

CONCLUSIONS

We conducted a controlled trial of an eco-driving assistance system in a driving simulator, comparing it with normal driving and eco-driving. The simulator data was used to evaluate fuel and energy consumption for conventional, hybrid electric, and battery electric powertrains. This revealed that the eco-driving assistance system leads to reductions in energy usage, with proportionally greater reductions in the hybrid and electric cases, and similar improvements for both male and female drivers. Eco-driving outperformed the assistance system for the electric powertrain, suggesting that it could be improved by designing specifically for electric vehicles. We also observed that the hybrid powertrain had lower fuel consumption than the conventional one in all trials, regardless of whether eco-driving was being used.

Therefore, eco-driving assistance and electrification may be viewed as complementary approaches to improving energy efficiency.

REFERENCES

Allison, C. K., Fleming, J. M., Yan, X., Lot, R., & Stanton, N. A. (2021). Adjusting the need for speed: assessment of a visual interface to reduce fuel use. *Ergonomics*, 64(3), 315–329.

Allison, C. K., Parnell, K. J., Brown, J. W., & Stanton, N. A. (2017). Modeling the real world using STISIM Drive simulation software: A study contrasting high and low locality simulations. In: *International Conference on Applied Human Factors and Ergonomics*, pages 906–915. Springer.

Barkenbus, J. N. (2010). Eco-driving: An overlooked climate change initiative. *Energy Policy*, 38(2), 762–769.

Beusen, B., Broekx, S., Denys, T., Beckx, C., Degraeuwe, B., Gijsbers, M., Scheepers, K., Govaerts, L., Torfs, R. & Panis, L. I. (2009). Using on-board logging devices to study the longer-term impact of an eco-driving course. *Transportation research part D: transport and environment*, 14(7), 514–520.

Cestac, J., Paran, F., & Delhomme, P. (2011). Young driver's sensation seeking, subjective norms, and perceived behavioral control and their roles in predicting speeding intention: How risk-taking motivations evolve with gender and driving experience. *Safety Science*, 49(3), 424–432.

Dib, W., Chasse, A., Moulin, P., Sciarretta, A., & Corde, G. (2014). Optimal energy management for an electric vehicle in eco-driving applications. *Control Engineering Practice*, 29, 299–307.

Fleming, J., Yan, X., Allison, C., Stanton, N., & Lot, R. (2018, October). Driver modeling and implementation of a fuel-saving ADAS. In *2018 IEEE International Conference on Systems, Man, and Cybernetics (SMC)* (pp. 1233–1238). IEEE.

Franke, T., Arend, M. G., McIlroy, R. C., & Stanton, N. A. (2016). Eco-driving in hybrid electric vehicles – Exploring challenges for user-energy interaction. *Applied Ergonomics*, 55, 33–45.

Hu, J., Shao, Y., Sun, Z., Wang, M., Bared, J., & Huang, P. (2016). Integrated optimal eco-driving on rolling terrain for hybrid electric vehicle with vehicle-infrastructure communication. *Transportation Research Part C: Emerging Technologies*, 68, 228–244.

Kamal, M. A. S., Mukai, M., Murata, J., & Kawabe, T. (2010). On board eco-driving system for varying road-traffic environments using model predictive control. In: *IEEE International Conference on Control Applications*, pages 1636–1641. IEEE.

Knowles, M., Scott, H., & Baglee, D. (2012). The effect of driving style on electric vehicle performance, economy and perception. *International Journal of Electric and Hybrid Vehicles*, 4(3), 228–247.

McIlroy, R. C., Stanton, N. A., Godwin, L., & Wood, A. P. (2016). Encouraging eco-driving with visual, auditory, and vibrotactile stimuli. *IEEE Trans. on Human-Machine Systems*, 47(5), 661–672.

Meschtscherjakov, A., Wilfinger, D., Scherndl, T., & Tscheligi, M. (2009). Acceptance of future persuasive in-car interfaces towards a more economic driving behaviour. In: *Proceedings of the 1st International Conference on Automotive User Interfaces and Interactive Vehicular Applications*, pages 81–88. ACM.

Neumann, I., Franke, T., Cocron, P., Bühler, F., & Krems, J. F. (2015). Eco-driving strategies in battery electric vehicle use – How do drivers adapt over time? *IET Intelligent Transport Systems*, 9(7), 746–753.

Paganelli, G., Ercole, G., Brahma, A., Guezennec, Y., & Rizzoni, G. (2001). General supervisory control policy for the energy optimization of charge-sustaining hybrid electric vehicles. *JSAE Review*, *22*(4), 511–518.

Tate, E., Harpster, M. O., & Savagian, P. J. (2008). The electrification of the automobile: From conventional hybrid, to plug-in hybrids, to extended-range electric vehicles. *SAE International Journal of Passenger Cars-Electronic and Electrical Systems*, *1*(2008-01-0458), 156–166.

Wu, C., Zhao, G., & Ou, B. (2011). A fuel economy optimization system with applications in vehicles with human drivers and autonomous vehicles. *Transportation Research Part D: Transport and Environment*, *16*(7), 515–524.

Yan, X., Fleming, J., & Lot, R. (2017). Modelling and energy management of parallel hybrid electric vehicle with air conditioning system. In: *2017 IEEE Vehicle Power and Propulsion Conference*, pages 1–5. IEEE.

Yan, X., Allison, C. K., Fleming, J. M., Stanton, N. A., & Lot, R. (2021). The benefit of assisted and unassisted eco-driving for electrified powertrains. *IEEE Transactions on Human-Machine Systems*, *51*(4), 403–407.

10 Predictive Eco-Driving Assistance on the Road

INTRODUCTION

Since 1990, significant progress has been made on decarbonisation of economies within the European Union (EU), with total EU carbon dioxide (CO_2) emissions 23.5% lower in 2017 than they were in 1990 (UNFCCC). While emissions from sectors such as electricity production, agriculture, and industry have decreased significantly from 1990 levels (by 29.9%, 19.2%, and 36.7% respectively), emissions from road transport have increased by 22.8% in the same period. As a sector, road transport accounts for 20.7% of total CO_2 emissions across the EU, so significant reductions in road transport emissions are needed to meet future decarbonisation targets.

Eco-driving is the reduction of fuel usage of a road vehicle by driving it in an energy-efficient way and has been described as "low-hanging fruit" that can provide a 5–10% reduction in fuel usage and resulting CO_2 emissions without requiring changes to existing vehicle drivetrains (Vandenbergh et al., 2007). This requires driving behaviours such as gentle acceleration, avoiding braking by anticipating traffic flow, and early upshifting of gears in cars with manual transmissions. These changes in behaviour may be achieved either by driver training or by incorporating driver assistance devices into the vehicle (Barkenbus, 2010). Many such devices have been suggested ranging from simple displays that give feedback to the driver about their fuel economy, to more sophisticated haptic and combined auditory-visual interfaces (Jamson et al., 2015). Further innovations include the incorporation of feedback on drivers' past actions and feedforward advice to make context-dependent recommendations, but as yet most such devices are unable to detect the presence of a vehicle ahead (Orfila et al., 2015). As well as improving fuel efficiency beyond that achievable by drivers alone, these assistance systems can induce learning effects in drivers to further promote more economical driving (Daun et al., 2013). This is particularly promising for long-term CO_2 reduction efforts, as eco-driving behaviours are reported as easy to adopt by drivers (Delhomme et al., 2013), but difficult to maintain long-term without regular feedback as drivers fall back into old habits (Pampel et al., 2018).

Dynamic eco-driving systems, in which advice is communicated to the driver in real-time based on sensor and infrastructure information, are demonstrated to give reductions in fuel consumption of 10–20% in simulation studies of highway driving situations (Barth & Boriboonsomsin, 2009), with the greatest reductions occurring in congested traffic. One such system was prototyped within the EU EcoDriver project, with a precise polynomial model of fuel consumption minimised using dynamic programming to develop an optimal speed profile and the resulting algorithm tested on a real vehicle in a restricted test-track situation (Cheng et al., 2013).

These systems are also useful for electric vehicles where they can be used to extend the possible driving range (Madhusudhanan, 2019). Pre-computed optimal speed profiles can be displayed to the driver via the speedometer (Lin et al., 2014). In surveys of car drivers, eco-driving assistance systems are rated as useful, with drivers welcoming their deployment in general (Trommer & Höltl, 2012). The sensing, infrastructure communication, and optimisation required for such systems may also be useful in adaptive cruise control applications designed to reduce fuel usage (Asadi & Vahidi, 2010).

In determining the eco-driving behaviour suitable for a given driving situation, optimal control has been identified as a useful framework, with analytical solutions possible in some simple cases and numerical solution methods being applicable to more general problems (Sciarretta et al., 2015). For real-time eco-driving assistance, model predictive control (MPC) has been demonstrated in traffic simulations to give fuel-efficiency benefits by solving an optimal control problem that incorporates information on the positions of other vehicles and prediction of the acceleration behaviour of the preceding vehicle based on, for example, traffic signals (Kamal et al., 2010). More recently researchers have also considered the effect of road curvature on eco-driving solutions (Ding & Jin, 2018), accounting for limits on the allowable lateral acceleration in curves. The optimal speed is constant while traversing a circular curve, but varies in a complex way on clothoid curves and on typical roads with transitions between straight, clothoid, and circular sections.

A major concern in the development of any eco-driving assistance system is user acceptance (Meschtscherjakov et al., 2009). To this end, the system should respect the driver's preferences on relevant quantities such as lateral and longitudinal accelerations, cruising speeds, and following distances. The possibility of making driver assistance reflect driver behaviour has already been suggested for adaptive cruise control, in which the system may be trained by drivers to reflect their preferences on speed and following distance (Simonelli et al., 2009). In this context of improving driver assistance by consideration of driver preferences, recent works by the authors have developed parametric models of driver behaviour suitable for safety (Fleming et al., 2019) and eco-driving assistance applications (Fleming et al., 2020). Having developed this theory, the present work applies a similar model to a prototype eco-driving assistance system.

This chapter concerns the final prototype of the eco-driving assistance system developed by the authors at the University of Southampton, UK, which accounts for both the motion of the preceding vehicle and upcoming road curvature to suggest actions to the driver that minimise fuel consumption (Fleming et al., 2021). The fuel-efficient velocity profile is determined by solving a nonlinear MPC optimisation numerically in real-time, exploiting measurements from GPS, and long-range automotive radar to provide knowledge of upcoming road geometry and traffic. This assistance system has several improvements and innovations compared to those in the existing literature:

- As the system considers both path curvature of the driving route and motion of any leading vehicles, it is general-purpose system suggesting fuel-efficient behaviour in real-time in a variety of scenarios such as urban and rural driving and both high and low traffic densities.

- Driver preferences on longitudinal and lateral acceleration, cruising speed, and vehicle spacing are incorporated, such that the system coaches the driver into fuel-efficient behaviour consistent with these preferences.
- The system contains an adjustable trade-off parameter α that can be personalised to either give more emphasis to fuel economy, which requires coasting down and gentle accelerations, or naturalistic driving, which admits higher accelerations and more closely follows driver preferences on speed and following distances.

In this chapter, we describe the architecture and development of the system as it was implemented in on-road testing, following a previous conference publication detailing the system as implemented in a driving simulator (Fleming et al., 2018a). We then provide results from a repeated-measures study of 36 participants driving in the simulator, designed to assess reductions in fuel usage as calculated from a detailed quasi-static model of the simulator vehicle powertrain. A repeated-measures analysis of variance (ANOVA) that includes average speed in each test condition as covariates allows changes in fuel usage to be compared to those that would be expected given any changes in average speed when using the system so that we can distinguish between reductions in fuel usage caused by participants driving more slowly, and those from other causes such as avoidance of braking. We then describe on-road testing to assess technical feasibility of the system. Applying the system to a real-world vehicle required the development of a suitable fuel consumption model, which was validated using measurements from the vehicle CANbus. Finally, we provide experimental results on the availability and reliability of sensor data and provide qualitative analysis of the system performance in practice.

SYSTEM ARCHITECTURE

Schematically, the system is divided into "Perception", "Decision", and "Action" layers as illustrated in Figure 10.1. The Perception layer consists of a GPS unit, front-mounted doppler radar sensor, and the vehicle ECU which can provide data on vehicle speed, gear, and instantaneous fuel consumption. The Decision layer is a receding-horizon MPC scheme that attempts to minimise a weighted sum of fuel consumption, acceleration, speed, and following distance objectives. This is informed by models of vehicle fuel consumption, driver preferences, and road curvature which are described in detail in later sections. The action layer consists of a visual human-machine interface (HMI) taking the form of a green and orange "eco-band" overlaid on a vehicle speedometer, inspired by "eco-speedometer" designs rated highly in usefulness and user acceptance in previous studies (Meschtscherjakov et al., 2009).

Considering the in-vehicle prototype developed by the authors, communication between the Perception and Decision layers occurs over the vehicle CANbus, with the Decision and Action layers implemented on a tablet PC in the vehicle cabin for the on-road prototype, while communication between the Decision and Action layers was implemented using TCP/IP. For simulator testing, the Perception layer was replaced with updates from the simulation software. A PC present in the simulator

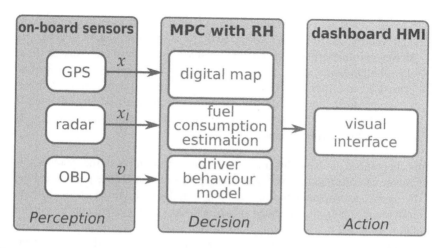

FIGURE 10.1 System architecture.

laboratory was used to run both the simulation software and the Decision layer, with the Action layer on a tablet PC in the vehicle cabin.

PERCEPTION LAYER

GPS-Based Localisation

To account for upcoming road curvature, the vehicle must be located on an internal map of the route which can be accomplished using GPS measurements of the vehicle's latitude and longitude. This internal map is specified as a sequence of x-y points representing the route in a local coordinate system, allowing translation of the measured latitude and longitude into a distance x along the route appearing in a curvature function $\kappa(x)$, which is used in the Decision layer. In the on-road prototype, the vehicle was equipped with a Racelogic VBOX 3i GPS unit and data logger, which outputs data continuously at a rate of 100 Hz to the vehicle CANbus.

Each time an updated latitude-longitude pair is received from the GPS, this is converted to the local coordinate system, denoted as a vector $\underline{y}$ and the nearest two points on the route, denoted $\underline{y}_k$ and $\underline{y}_{k+1}$, are identified. If the distances of $\underline{y}_k$, $\underline{y}_{k+1}$ along the route are x_k and x_{k+1} respectively, then the current distance x may be approximated by the interpolation:

$$x = x_k + \sigma(x_{k+1} - x_k) \tag{10.1}$$

where $0 \le \sigma < 1$ measures the progress of the vehicle in travelling from $\underline{y}_k$ to $\underline{y}_{k+1}$ and may be given by:

$$\sigma = \frac{\left(\underline{y}_{k+1} - \underline{y}_k\right)}{\|\underline{y}_{k+1} - \underline{y}_k\|} \cdot \frac{\left(\underline{y} - \underline{y}_k\right)}{\|\underline{y}_{k+1} - \underline{y}_k\|}$$

This ensures that the current value of x, used in the Decision layer, increases smoothly as the vehicle travels along the route.

If the road positions y are accurate and the vehicle remains on the road, the position error of the localisation method is less than or equal to the position error of the GPS data. Because in the worst case, any position error from the GPS is aligned with the road direction so that the error in x is equal to it. Noting that the VBOX 3i has a horizontal position error of up to 1.2 m (root-mean-square), the time taken to travel this distance at any speed is great enough to require deceleration before a curve or road feature will be small compared to the time constants of the vehicle dynamics. We, therefore, believe that any inaccuracy should have only a minor effect on the functioning of the assistance system. Nonetheless, if more accuracy were desired, measurements of integrated vehicle speed (available from the CANbus/ECU), or of road slope (which is known from mapping data), could be combined with the location x using sensor fusion techniques.

LONG-RANGE RADAR SENSING

The test vehicle is equipped with a front-mounted TRW AC-10 long-range radar. This communicates via the CANbus and is capable of simultaneous tracking of up to 8 objects in its standard configuration, with a field of view of 12° and an operating range of 2 to 200 m. The radar performs basic signal processing internally to track detected objects, and measurements such as the range, relative velocity, and signal strength are output to the CANbus in real-time, along with several flags to indicate whether a tracked object is in the same lane as the current vehicle, and whether it is moving in the same direction, opposite direction, or is stationary. The range measurements are accurate to within ±0.1 m and speed measurements accurate to ±0.1 m/s for typical ranges observed during vehicle following.

For implementation of the eco-driving assistance system, these flags were used to perform basic filtering to identify which (if any) of the currently tracked objects correspond to a lead vehicle. This was done by retaining only those tracks in the current lane and choosing the remaining track with the greatest signal strength. The range and relative velocity of this lead vehicle was then sent to the Decision layer at a rate of 20 Hz.

VEHICLE ECU

Real-time data on vehicle speed, revolutions per minute (RPM), instantaneous fuel consumption, and many other variables are transmitted from the ECU to the vehicle CANbus at a rate of 100 Hz, where they may be read by the Decision layer. In the on-road prototype described in this chapter, the RPM and speed measurements were also forwarded to the visual HMI, which doubled as a working speedometer and tachometer. This data from the ECU was acquired and combined by the Perception layer software with the GPS and radar data before being forwarded to the Decision layer at a steady rate of 20 Hz, implying that all filtering and data processing was carried out in less than 50 ms.

TABLE 10.1
Polynomial Coefficients of Selected Fuel Model

a_{ij}	i = 0	i = 1	i = 2	i = 3	i = 4
j = 0	3.28e-01	−2.71e-01	1.81e-01	−2.74e-02	1.40e-03
j = 1	−2.31e-02	7.72e-02	−1.96e-02	1.40e-03	–
j = 2	1.91e-03	8.50e-04	5.84e-04	–	–
j = 3	−1.21e-04	−5.41e-05	–	–	–
j = 4	2.76e-06	–	–	–	–

DECISION LAYER

FUEL CONSUMPTION MODEL

To determine the acceleration and braking behaviour minimising fuel usage, a mathematical model of instantaneous fuel consumption is required. Rather than explicitly considering gearing, which would complicate the optimisation, we model the instantaneous fuel consumption as a function $L_f(F, v)$ of the force F and velocity v at the wheels of the vehicle. This introduces some error as any particular combination of force and velocity may be achieved in different gears, leading to different operating points for the engine and different resulting fuel consumptions. However, it also greatly simplifies the implementation of the optimisation which does not need to consider the gear ratio. Hence the mass m_f of fuel consumed during a journey taking time T is given by:

$$m_f = \int_0^T L_f(F, v)\, dt \qquad (10.2)$$

If it is assumed to have a known standard form, such as a polynomial, the function $L_f(F, v)$ may be found via regression using data from the vehicle, giving coefficients as shown in Table 10.1 for a 4th order polynominal model. Note that as this fuel consumption model is static and not dynamic, it necessarily cannot model dynamic effects such as the effect of engine temperature, and hence should be considered valid only in normal operating conditions when the engine has warmed up.

During a test drive, the authors collected data on instantaneous fuel consumption as reported by the ECU and collected from the vehicle CANbus. This test drive was performed on a 21km route around Southampton, UK, and contained urban, rural, and motorway sections driven at an average of 53 kilometres per hour. Further info about the route used may be found in the following section on the performed simulator study. This data was split into "training" and "test" subsets in proportions of 80%/20%. Polynomial models of fuel consumption of differing orders were then fit using linear regression. These models had the form:

$$L_f(F, v) = \sum_i \sum_j a_{ij} F^i v^j \qquad (10.3)$$

with the range of i and j in each case chosen to retain terms up to the order of the model. The resulting R^2 values along with the root mean square prediction error

Predictive Eco-Driving Assistance on the Road

TABLE 10.2
Comparison of Polynomial Fuel Consumption Models

Degree	R^2 (train. data)	RMSE (cross-valid.)
1	0.716	0.4768
2	0.944	0.2129
3	0.945	0.2124
4	0.950	0.2046
5	0.950	0.2064
6	0.951	0.2276

(RMSE) estimated by 10-fold cross-validation on the training dataset is given in Table 10.2. The fourth-order model has the minimum cross-validation error, with higher-order models showing evidence of overfitting. This model was selected as the fuel consumption model to be used in the on-road prototyping and the RMSE was also calculated over the test data as 0.2047 ml/s. A contour map of this model is shown in Figure 10.2, and a comparison of the actual and predicted model values over the test data is provided in Figure 10.3.

DRIVER PREFERENCE MODEL

As the assistance system is designed to manage variations in speed due to cornering and vehicle-following, it should model driver preferences in these situations and contain tunable parameters for typical accelerations, following distances, and lateral

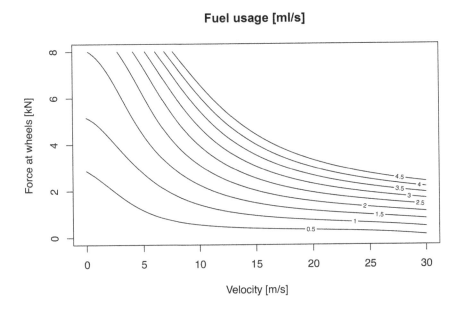

FIGURE 10.2 Contour map of fourth-order fuel consumption model.

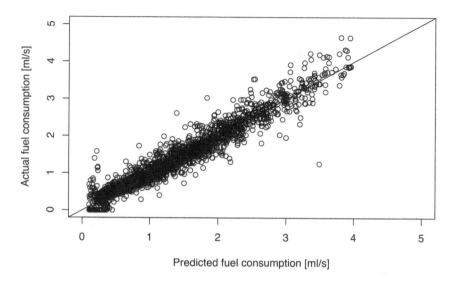

FIGURE 10.3 Predicted vs. actual fuel consumption, test data.

acceleration limits. In terms of the optimisation, this is accomplished by including a combination of penalty functions that should be minimised to achieve natural car-following behaviour and inequality constraints that limit acceleration and vehicle speed in curves. The former may be expressed as the integral:

$$S_d = \int_0^T L_d(s,v,a)\,dt \tag{10.4}$$

in which we have,

$$L_d(s,v,a) = a^2 + \frac{4}{v_d}(v-v_d)^2 + \frac{(1-s/s_d)^2}{(s/s_d)^2+1} \tag{10.5}$$

where v and a are the vehicle velocity and acceleration, $s = x_l - x - l$ is the headway distance to a preceding vehicle with position x_l and length l, and $s_d = T_{min}v + s_{min}$ is a desired distance to the preceding vehicle which increases linearly with speed. Further details of the development of the cost function (10.5) may be found in (Fleming et al., 2018a). To limit the driver's maximum acceleration, we also constrain a as:

$$a \leq a_{max} \tag{10.6}$$

Moreover, a speed-dependent limit on lateral acceleration is applied when the vehicle is following a path of curvature κ, leading to a constraint on velocity:

$$v \leq \sqrt{\frac{\Gamma_{max}}{\kappa + \Delta\kappa_{max}}} \tag{10.7}$$

TABLE 10.3
Parameters for the Driver Model

Parameter	Value	Physical Interpretation
v_d	30 ms^{-1}	Free-flow vehicle speed
s_{min}	2 m	Minimum distance to lead vehicle
T_{min}	1.2 s	Minimum time headway to leading vehicle
a_{max}	5 m/s^2	Maximum longitudinal acceleration
Γ_{max}	5 m/s^2	Maximum lateral acceleration
$\Delta\kappa_{max}$	3 rad/km^{-1}	Curvature safety margin

This may be rearranged as:

$$(\kappa + \Delta\kappa_{max})v^2 \leq \Gamma_{max}$$

so that we may interpret the velocity bound as the driver applying an upper limit to the lateral acceleration $\kappa v^2 < \Gamma_{max}$ while allowing for a possible error $\Delta\kappa_{max}$ in their estimation of the curvature of the vehicle's path.

The six parameters a_{max}, v_d, s_{min}, T_{min}, Γ_{max}, and $\Delta\kappa_{max}$ are summarised in Table 10.3 along with some typical values. These parameters were designed to be a subset of those in existing models for cornering and car-following. In the present work, these values were chosen based on observed values in real-world driving, but in a practical system, they could be adapted to individual drivers using naturalistic data (Fleming et al., 2018b).

PREDICTIVE OPTIMISATION OF VEHICLE SPEED

The longitudinal motion of the vehicle is modelled by considering it as a mass subject to forces due to the drivetrain, aerodynamic drag, rolling resistance, and road slope. The dynamics may then be expressed as:

$$\dot{x} = v \qquad (10.8)$$

$$m\dot{v} = F - \frac{1}{2}\rho_a C_d A v^2 - mg(\sin\theta + C_{rr}\cos\theta)$$

where C_d and C_{rr} denote coefficients of drag and rolling resistance and we have assumed $v \geq 0$ to simplify those terms. For prediction of the preceding (lead) vehicle position, its dynamics are also included in the optimisation problem as:

$$\dot{x}_l = v_l \qquad (10.9)$$

in which the leader velocity v_l is assumed constant over the prediction interval. To state the optimisation problem, we must also introduce initial conditions for these variables as:

$$x(0) = x_0, \quad v(0) = v_0, \quad x_l(0) = x_{l0} \tag{10.10}$$

where x_0, v_0, and x_{l0} are provided by the Perception layer.

Introducing a parameter α to trade-off fuel usage with the driver's preferences, and a shortened time horizon $T_h < T$, the full optimisation problem may now be expressed as:

$$\underset{x(t),v(t),F(t)}{\text{minimise}} \int_0^{T_h} \left[L_d(s,v,\dot{v}) + \alpha L_f(F,v) \right] dt \tag{10.11}$$

subject to (10.6 – 10.10).

The effect of the parameter α is to trade-off the driver's preferences and the minimisation of fuel consumption (Fleming et al., 2020). Greater reductions in fuel consumption are possible if following distances and velocities are allowed to vary from their nominal values, so as α is increased the fuel consumption of the computed speed profile will decrease. However, large values may lead to behaviour that is unacceptable to a human driver. The reduction of the time interval from T to T_h modifies the later parts of the solution as t approaches T_h. This may be mitigated by including a terminal cost designed to penalise deviations from a target average speed as described in (Fleming et al., 2018a). This optimisation is carried out in a receding-horizon manner, using updated values for the initial conditions (10.10) and providing feedback to updated measurements. In the on-road prototype, the optimisation was implemented in the ACADO toolkit (Houska et al., 2011), exploiting its capability for nonlinear receding-horizon control. Real-time data from the Perception layer was used to set the initial conditions, and the prediction horizon $T_h = 60$ s. This problem was solved in real-time at a rate of 2 Hz, with a mean computation time of 0.23 s. We made no attempt to shorten this computation time, but it is likely it could be reduced further if code generation techniques were employed. The resulting optimised velocity at $t = 10$ seconds into the prediction horizon was sent to the Action layer for display, to "coach" the driver into following the optimised trajectory.

ACTION LAYER

VISUAL INTERFACE

The result of the predictive eco-driving optimisation was displayed to the driver using a simple visual HMI, consisting of a green and orange "eco-band" overlayed on the vehicle speedometer. This design was chosen from competing alternatives in a collaborative workshop in which interface proposals were rated in several categories using the "Design with Intent" methodology (Allison et al., 2018). When in use, the green section extends from zero speed up to the recommended speed received from the Decision layer. The orange region extends above this, allowing some margin for

Predictive Eco-Driving Assistance on the Road

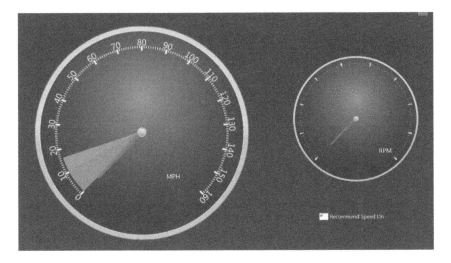

FIGURE 10.4 The visual HMI, showing recommended speeds to the driver.

error, and has a width chosen to correspond to the typical root-mean-square deviation of speed in normal driving data when cruising at the speed limit. When in use, the interface updates in real-time at a rate of 60 Hz. As data is received from the Decision layer, the recommended speed changes and the interface smoothly interpolates between values in order to coach the driver into following the optimal speed profile. For the simulator and on-road prototypes, this visual interface was developed in C# using Windows Presentation Foundation as a graphical library. The resulting application is executed on a Microsoft Surface Pro tablet, placed behind the steering wheel of the car to replace the speedometer, and is shown in Figure 10.4.

SIMULATOR TESTING

TEST PROCEDURE

For initial development of the eco-driving assistance system, and to allow for a controlled trial of the system on different drivers, the authors developed an initial prototype in a fixed-base driving simulator. As shown in Figure 10.5, this consisted of a 2015 Land Rover Discovery Sport with three large projector screens at the front, an additional screen to the rear, and LCD displays in the side mirrors, to simulate the view from a real vehicle. Engine sounds are simulated using the vehicle's internal speakers. STISIM Drive was used as the simulation software to display the road environment. Drivers carry out the driving task by using the foot pedals and steering wheel within the vehicle cabin, following a simulated 21 km route around Southampton, UK, with urban, rural, and motorway (highway) sections. This is the same route as was used for the determination of the fuel consumption model. The simulation software recorded time-series data including vehicle speed and throttle and brake inputs, which was then used to estimate fuel usage for each study participant based on a detailed quasi-static model of the vehicle powertrain implemented in Simulink, which considered an engine

FIGURE 10.5 The fixed-base driving simulator used for trials.

map of fuel consumption losses due to the vehicle transmission, aerodynamic drag, and rolling resistance. These drivers completed the route three times under different conditions: normal driving ("Control"), unassisted eco-driving ("Eco"), and eco-driving with the assistance system enabled ("Band"):

- "Control" condition: Participants were instructed to drive as they "usually would", with no other specific instructions. The eco-driving assistance system was turned off, and the tablet PC in the vehicle cabin showed only a speedometer and tachometer.
- "Eco" condition: Participants were told to drive "as fuel-efficiently as possible", after being instructed in eco-driving behaviours such as gentle acceleration, avoidance of heavy braking, and maintaining a steady speed. The assistance system was turned off.
- "Band" condition: The eco-driving assistance band was turned on. Participants were told that this speed recommendation would help them conserve fuel if followed. They were encouraged to use the interface as long as it did not interfere with other driving tasks.

To lessen the impact of possible order effects, the study was fully counterbalanced in that the order in which drivers were exposed to each condition was varied such that an equal number of participants carried out each of the six possible permutations. A total of 36 participants took part in the study, with an equal number of males and females. They were aged between 18 and 71 years (mean: 28.9 years, standard deviation: 12.82). All participants were either working in or resident in Southampton and licensed to drive in the UK, so may be expected to be familiar with the location and roads used for the study, although they may have not previously driven the simulated route. Participants completed two practice drives prior to the main study, one with the assistance system enabled. The route included simulated traffic, which was identical on each repetition of the test. Traffic density was based on early afternoon levels with no jams but some stop-and-go behaviour due to traffic signals, with a stable level of traffic flow in the urban situation (density approx. 10–20 vehicles per mile) and free-flowing traffic in the motorway and rural situations (density <10 vehicles per lane per mile).

RESULTS

A boxplot of the distribution of fuel consumption for the different conditions and road types is given in Figure 10.6, showing that the median consumption was highest in the Control condition and lowest in the Eco condition for each road type. From physical considerations of aerodynamic drag, we expect that lower average speeds over the route will typically lead to lower values of fuel consumption. Figure 10.7 shows that this is the case. To estimate the effect of the Eco and Band conditions while controlling for the effects of differing average speed, we fit a repeated-measures ANOVA (GLM repeated measures) model to the fuel consumption data, with Condition and Road Type (Urban, Rural, and Motorway) as within-subjects factors and the average speed in each test condition included as covariates. The data met the standard assumptions (independence, normality of residuals, homogeneity of variances) for use of a linear model. Sphericity was checked using Mauchly's test, with results indicating that the assumption of sphericity was not violated for Condition ($\chi^2(2) = 4.31, p = 0.116$), Road Type ($\chi^2(2) = 3.38, p = 0.188$), or their interaction ($\chi^2(9) = 13.9, p = 0.126$).

Considering a standard significance level of $\alpha = 0.05$, analysis indicated that both road type ($F(2,64) = 4.57$, $p = 0.011$, $\eta_p^2 = 0.028$) and condition ($F(2,64) = 15.05$, $p < 0.001$, $\eta_p^2 = 0.088$) had significant effects on fuel consumption, while their interaction ($F(4,128) = 2.06$, $p = 0.085$, $\eta_p^2 = 0.026$) did not. Considering the question of whether the Band condition had different fuel consumption than the Control condition, post hoc analysis using Tukey's test estimated a 0.249 l/100 km improvement

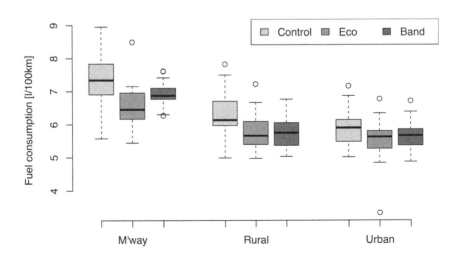

FIGURE 10.6 Fuel consumption in simulator testing.

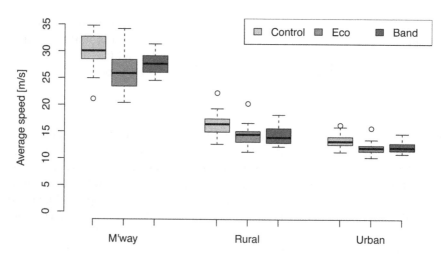

FIGURE 10.7 Average speeds in simulator testing.

in fuel consumption in the Band condition when the effect of average speed is controlled for (95% CI: [0.084, 0.415] l/100 km, $p = 0.001$). Noting that the mean fuel consumption in the Control and Band conditions was 6.501 l/100 km and 6.105 l/100 km respectively, the overall improvement in fuel consumption in the Band case was 6.09%, which is 3.96% greater than expected from the reduction in average speed alone. There was no significant difference between the fuel consumption in the Band and Eco-driving conditions once average speed was controlled for. Although no significant differences were observed from the Band case, drivers reduced fuel consumption further during unassisted eco-driving by travelling more slowly in aggregate on the motorway. Participants' average speed in the motorway part of the test showed a mean of 26.1 m/s with a standard deviation of 3.11 m/s in the Eco condition, while for the Band condition the mean was 27.6 m/s and standard deviation 1.92 m/s. Hence use of the assistance system was associated with an increase in motorway average speed but a decrease in its variance. Interestingly, this led to a decreased number of drivers with average speeds exceeding the legal speed limit of 31.3 m/s (15 during "Control", which had mean 30.2 m/s and standard deviation 3.05 m/s, 1 in "Eco", and 0 in "Band"). Travel time on the motorway was also improved as a result of this greater average speed, with a mean travel time of 255 s for the 7 km motorway section in the Band condition, 6.48% lower than the mean travel time of 272 s in the Eco condition.

ON-ROAD TESTING

Test Procedure

To evaluate the practical feasibility of the system, the authors developed a prototype system on the instrumented vehicle shown in Figure 10.8. The vehicle is a 2004 Fiat Stilo with a 6-speed automatic transmission and was equipped with the radar and

Predictive Eco-Driving Assistance on the Road 219

FIGURE 10.8 The instrumented vehicle used for testing.

GPS units used for the study as well as a tablet PC to run the Decision and Action layers of the system.

Two routes were used for testing of the prototype. These are shown in Figures 10.9 and 10.10 and were designed to test the response of the system in vehicle-following and cornering situations respectively. During the car-following test, the driver followed another vehicle on the same route, while the cornering test was carried out when the

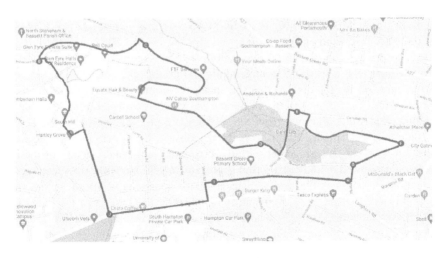

FIGURE 10.9 Following route. Map data: © 2021 Google.

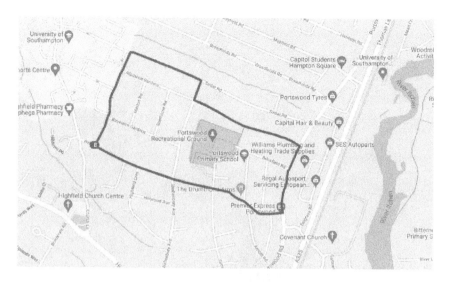

FIGURE 10.10 Cornering route. Map data: © 2021 Google.

route was mostly free of other traffic. As the fuel-saving potential of the system was assessed separately in the driving simulator, this on-road testing was limited to a technical evaluation of the system, concentrating in particular on the accuracy of a fuel consumption model based only on velocity and wheel force, and any reduction in performance caused by the need to obtain real-world data from sensors rather than from a driving simulation. These routes were driven three times in each case.

RESULTS

Figure 10.11 shows the actual vehicle speed and speed recommended by the system during an 80s section of the cornering test which contained two consecutive curves. The steering angle and number of visible GPS satellites is also shown. It is clear that the system recommends a lower speed starting around 20s before the curve at 360s. The driver, who was instructed to follow the speed recommendation for this portion of the test, uses this recommendation to gradually reduce speed approaching the curve. It is also notable that the speed recommendation begins to increase approximately 10s before the end of the curve, which is expected recalling that the recommended value is drawn from the speed at 10s into the prediction horizon of the solution of (11). Figure 10.12 shows the response of the system for a short section of the car-following test, together with the estimated range and velocity of the lead vehicle as measured by the radar. From approximately 850 to 900s, the speed recommendation increases as the lead vehicle gradually increases its speed, before radar tracking is lost intermittently from approximately 905s to 925s. Radar tracking of the lead vehicle is recovered at 923s, where it has come to a halt and the speed recommendation also reflects this fact. This was a common occurrence during the test, with the system often recommending slowing down when it detected a stopped or slow vehicle ahead. Tables 10.4 and 10.5 respectively show the availability of measurements from

Predictive Eco-Driving Assistance on the Road

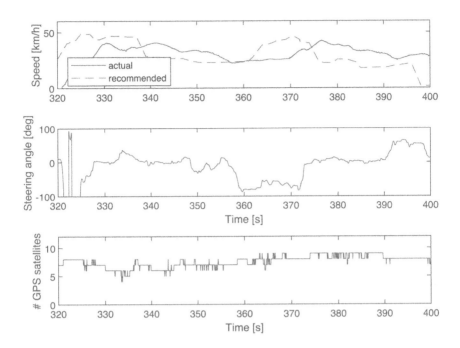

FIGURE 10.11 System response in cornering test.

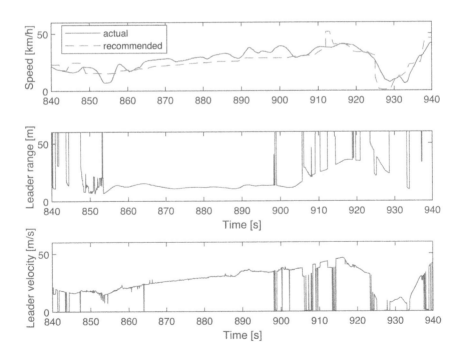

FIGURE 10.12 System response in following test.

TABLE 10.4
GPS Coverage during Test

# Satellites Visible	% of Time Following	Cornering
≥3	99.6%	98.9%
≥4	98.2%	98.4%
≥5	95.4%	96.1%
≥6	77.3%	90.3%
≥7	10.0%	76.2%
≥8	0%	55.1%
≥9	0%	28.1%
≥10	0%	0.9%

TABLE 10.5
Radar Tracking during Test

# Tracked Objects	% of Time Following	Cornering
≥1	94.3%	92.2%
≥2	69.9%	67.3%
≥3	55.1%	35.4%
≥4	27.0%	16.3%
≥5	7.8%	7.7%
≥6	2.3%	3.5%
≥7	0.7%	1.2%
≥8	0.3%	0.0%

the GPS and radar during the tests. For the cornering test, up-to-date GPS measurements were available 98.4% of the time (corresponding to when there were at least 4 satellites visible), but in the following test, the leader was successfully tracked only 67.3% of the time. This was due to a combination of intermittent errors in tracking such as those in Figure 10.12, typically caused by other metallic objects, and longer instances in which the lead vehicle was outside of the operating angle of the radar.

DISCUSSION

Qualitatively, the system behaved as expected during the on-road tests, though with some shortcomings in terms of the reliability of the available sensor data with availability of accurate radar measurements only 67.3% in a following situation. This was mostly due to the limited field-of-view of the radar, which may lose track of a lead vehicle if it moves out of this range during cornering, for example. This could perhaps be overcome by including leader position and velocity estimates from other sensors such as stereo cameras. These sensors would be available on vehicles with adaptive cruise control (ACC) systems or autonomous vehicles, for which this kind of eco-driving assistance system could be integrated into the ACC or autonomous driving software, replacing the Action layer of the present system by actuation that directly controls the speed of the vehicle to improve energy efficiency. Although an attractive concept, solving the Decision layer optimisation at a rate of 2 Hz would be inadequate for smooth control of the vehicle in such applications so would need to be increased. For driver assistance, there is also scope to replace the visual HMI within the Action layer of the system with other interfaces, such as a haptic accelerator pedal. To aid with electrification efforts, the model of fuel usage used could be replaced with one of electrical energy usage with only minor modifications to the Decision layer software. This would require negative forces F and negative consumption rates L_f to be considered when fitting the model, to allow for the possibility of regenerative braking.

The overall efficacy of the eco-driving assistance, relative to normal and unassisted eco-driving, can be evaluated from the results of the simulator testing. This showed no significant difference between the fuel consumption in the Band (assisted) and Eco (unassisted) conditions once the effect of average speed was controlled for. Both cases showed significant improvements in fuel economy over normal driving for a variety of road types. The higher traffic density encountered in the urban section of simulator testing implies that fuel savings in this section were due to differences in car-following and start-stop behaviour. In contrast, in the rural section of the drive, we expect that fuel savings were due to different behaviour when approaching curves, as the traffic was free-flowing in this section of the simulator test. The assistance system was designed to promote fuel-saving behaviour in both curves (for rural driving) and car-following (for urban/motorway situations) and managed to achieve this in both simulator and on-road testing. This reinforces the findings of previous studies showing that eco-driving solutions work in both congested start-stop situations (Barth & Boriboonsomsin, 2009) and curves (Ding & Jin, 2018) while additionally demonstrating that a general-purpose system can give benefits in both cases. Notably, in the unassisted Eco condition drivers reduced fuel consumption further by travelling more slowly on average on the motorway, with a negative effect on mean travel time. But this behaviour was avoided when drivers were assisted with the band, leading to a greater average speed overall and lower variances in the speeds of individual drivers. This is likely due to the system's design, with the desired speed setpoint v_d specified as part of the driver preference model. Average speed was also generally reduced in the Band case compared to the Control case, but much of this was a result of better compliance with the legal speed limit.

LIMITATIONS

The chosen minimum time headway parameter of $T_{\min} = 1.2$ is below the 2 s time headway recommended in most driving rulebooks (including the UK "Highway Code"), as it was chosen based on values of inter-vehicle distance observed in real-world driving. This could be raised to 2s if desired to improve safety, though we did not do so in the present study to ensure that the system did not recommend slowing down to increase vehicle spacings at typically observed headway distances.

We did not attempt to evaluate the driver distraction effects of the developed visual interface, as eye-tracking capability was not present within the driving simulator at the time of the study. Lateral position data were collected in the simulator relative to the vehicle position in the lane so that the frequency and severity of lane deviations could be considered in future works. We consider this outside of the scope of the present chapter, instead leaving it to a potential future publication investigating changes in participants' driving styles in the different conditions.

CONCLUSIONS

This chapter has covered the implementation and testing of an eco-driving assistance system designed to provide a real-time speed recommendation to a driver to assist them in saving fuel. The results of testing were generally positive, in particular:

- The simplified fuel-consumption model used in the system had an RMSE in testing of 0.205 ml/s and achieved an R^2 of 0.95.
- The optimisation was successfully implemented in an on-road prototype system and solved in less than 250 ms on average, allowing real-time implementation at 2 Hz.
- The improvement in fuel consumption in simulator testing was 6.09%, which is 3.96% greater than expected from reductions in average speed with the system on.
- Controlling for the effect of average speed, improvements in fuel economy of 0.25l/100 km were observed, similar to those for unassisted eco-driving.
- Motorway travel times were improved by 6.5% when using the assistance system versus unassisted eco-driving, while incidents of speeding decreased as a result of a reduced variance in speed.

REFERENCES

Allison, C., Stanton, N., Fleming, J., Yan, X., Goudarzi, F., & Lot, R. (2018). Inception, ideation and implementation: Developing interfaces to improve drivers' fuel efficiency. In: *Chartered Institute of Ergonomics and Human Factors (CIHEF)*.

Asadi, B., & Vahidi, A. (2010). Predictive cruise control: Utilizing upcoming traffic signal information for improving fuel economy and reducing trip time. *IEEE Transactions on Control Systems Technology, 19*(3), 707–714.

Barkenbus, J. N. (2010). Eco-driving: An overlooked climate change initiative. *Energy Policy, 38*(2), 762–769.

Barth, M., & Boriboonsomsin, K. (2009). Energy and emissions impacts of a freeway-based dynamic eco-driving system. *Transportation Research Part D: Transport and Environment, 14*(6), 400–410.

Cheng, Q., Nouveliere, L., & Orfila, O. (2013). A new eco-driving assistance system for a light vehicle: Energy management and speed optimization. In: *2013 IEEE Intelligent Vehicles Symposium (IV)*, pages 1434–1439. IEEE.

Daun, T. J., Braun, D. G., Frank, C., Haug, S., & Lienkamp, M. (2013). Evaluation of driving behavior and the efficacy of a predictive eco-driving assistance system for heavy commercial vehicles in a driving simulator experiment. In: *16th International IEEE Conference on Intelligent Transportation Systems (ITSC 2013)*, pages 2379–2386. IEEE.

Delhomme, P., Cristea, M., & Paran, F. (2013). Self-reported frequency and perceived difficulty of adopting eco-friendly driving behavior according to gender, age, and environmental concern. *Transportation Research Part D: Transport and Environment, 20*, 55–58.

Ding, F., & Jin, H. (2018). On the optimal speed profile for eco-driving on curved roads. *IEEE Transactions on Intelligent Transportation Systems, 19*(12), 4000–4010.

Fleming, J. M., Allison, C. K., Yan, X., Lot, R., & Stanton, N. A. (2019). Adaptive driver modelling in ADAS to improve user acceptance: A study using naturalistic data. *Safety Science, 119*, 76–83.

Fleming, J. M., Allison, C. K., Yan, X., Lot, R., & Stanton, N. A. (2019). Adaptive driver modelling in ADAS to improve user acceptance: A study using naturalistic data. *Safety Science, 119*, 76–83.

Fleming, J., Yan, X., & Lot, R. (2020). Incorporating driver preferences into eco-driving assistance systems using optimal control. *IEEE Transactions on Intelligent Transportation Systems, 22*(5), 2913–2922.

Fleming, J., Yan, X., Allison, C., Stanton, N., & Lot, R. (2021). Real-time predictive eco-driving assistance considering road geometry and long-range radar measurements. *IET Intelligent Transport Systems, 15*, 573–583.

Fleming, J., Yan, X., Allison, C., Stanton, N., & Lot, R. (2018, October). Driver modeling and implementation of a fuel-saving ADAS. In *2018 IEEE International Conference on Systems, Man, and Cybernetics (SMC)* (pp. 1233–1238). IEEE.

Houska, B., Ferreau, H. J., & Diehl, M. (2011). Acado toolkit – An open-source framework for automatic control and dynamic optimization. *Optimal Control Applications and Methods, 32*(3), 298–312.

Jamson, A. H., Hibberd, D. L., & Merat, N. (2015). Interface design considerations for an in-vehicle eco-driving assistance system. *Transportation Research Part C: Emerging Technologies, 58*, 642–656.

Kamal, M. A. S., Mukai, M., Murata, J., & Kawabe, T. (2010). On board eco-driving system for varying road-traffic environments using model predictive control. In: *2010 IEEE International Conference on Control Applications*, pages 1636–1641. IEEE.

Lin, X., Görges, D., & Liu, S. (2014). Eco-driving assistance system for electric vehicles based on speed profile optimization. In: *2014 IEEE Conference on Control Applications (CCA)*, pages 629–634. IEEE.

Madhusudhanan, A. K. (2019). A method to improve an electric vehicle's range: Efficient cruise control. *European Journal of Control, 48*, 83–96.

Meschtscherjakov, A., Wilfinger, D., Scherndl, T., & Tscheligi, M. (2009). Acceptance of future persuasive in-car interfaces towards a more economic driving behaviour. In: *Proceedings of the 1st International Conference on Automotive User Interfaces and Interactive Vehicular Applications*, pages 81–88. ACM.

Orfila, O., Saint Pierre, G., & Messias, M. (2015). An android based eco-driving assistance system to improve safety and efficiency of internal combustion engine passenger cars. *Transportation Research Part C: Emerging Technologies, 58*, 772–782.

Pampel, S. M., Jamson, S. L., Hibberd, D. L., & Barnard, Y. (2018). Old habits die hard? The fragility of eco-driving mental models and why green driving behaviour is difficult to sustain. *Transportation Research Part F: Traffic Psychology and Behaviour, 57*, 139–150.

Sciarretta, A., De Nunzio, G., & Ojeda, L. L. (2015). Optimal eco-driving control: Energy-efficient driving of road vehicles as an optimal control problem. *IEEE Control Systems Magazine, 35*(5), 71–90.

Simonelli, F., Bifulco, G. N., De Martinis, V., & Punzo, V. (2009). Human-like adaptive cruise control systems through a learning machine approach. In *Applications of Soft Computing* (pp. 240–249). Springer, Berlin, Heidelberg.

Trommer, S., & Höltl, A. (2012). Perceived usefulness of eco-driving assistance systems in Europe. *IET Intelligent Transport Systems*, *6*(2), 145–152.

UNFCCC. EEA greenhouse gas emissions and removals. https://www.eea.europa.eu/data-and-maps/data/data-viewers/greenhouse-gases-viewer. Accessed: 2019-09-09.

Vandenbergh, M. P., Barkenbus, J., & Gilligan, J. (2007). Individual carbon emissions: The low-hanging fruit. *UCLA L. Review*, *55*, 1701.

11 Designing for Eco-Driving
Guidelines for a More Fuel-Efficient Vehicle and Driver

INTRODUCTION

Throughout this book, we have investigated approaches to reduce fuel and energy usage and the related emissions, primarily carbon dioxide, that are associated with everyday driving. This chapter seeks to briefly summarise the key findings drawn from each of the chapters, to offer useful advice for those conducting future research and development or who seek to reduce their own personal emissions and fuel usage whilst driving.

Three immediate findings based upon the work presented are clear in that fuel use and emissions associated with everyday can be reduced by 1) Changing driver behaviour, 2) Developing more fuel-efficient vehicles, and 3) Using technology and in-vehicle interfaces to provide drivers clear, in the moment, guidance of how to minimise their fuel usage and emissions. It is hoped that readers can see that this final objective can only be achieved by cross-disciplinary partnership between human factors experts and automotive engineers. By promoting such collaboration, insight can be gain regarding not only human preferences and capabilities, but also insight regarding how these preferences can be realistically achieved.

CHAPTER SUMMARIES

Chapter 1 provided an overview of previous work, aligned from a human factors perspective, exploring fuel usage and emission reduction associated with driving behaviours. Previous research has consistently identified that the driver and the actions that a driver takes whilst driving have a significant impact on subsequent fuel use. This has created the notion of eco-driving, driving in a style to reduce fuel use and emissions. Eco-driving behaviours typically include gentle acceleration, minimisation of braking, and early gear changes in manual vehicles to minimise engine revolutions per minute (RPM). Numerous approaches have been used to try to encourage eco-driving behaviours when driving, primarily focusing on driver training and the use of in-vehicle interfaces to encourage eco-driving behaviours. When considering in-vehicle interfaces specifically, previous research has indicated that auditory feedback is generally disliked by users and is largely ineffective at changing behaviour. In contrast, both visual and haptic feedback are more readily accepted by drivers and have both been demonstrated to be effective at encouraging eco-driving. Despite this, safety concerns are apparent with the use of such interfaces, either due to fears relating to distraction effects, for visual interfaces, or fears related to being able to fully control the vehicle, in the case of haptic interfaces.

Chapter 2 sought to introduce readers to a dedicated framework to understand the design constraints of eco-driving, that of Cognitive Work Analysis (CWA). Constraints within this context are limits on drivers' behaviour which need to be respected to allow elements within a system, both human and non-human, to operate effectively. It can be seen from this chapter that CWA is an appropriate method for generating design constraints for eco-driving systems as it allows consideration of all actors within the system, across a variety of different likely future scenarios. By considering each step of the CWA process, this chapter acts as a guide for readers to be able to conduct their own analysis, with their own scenarios and design constraints. From the insight gained by this chapter regarding the constraints which drivers face, future work can be more accurately targeted to address specific scenarios. It is therefore argued that the use of a structured and evidence-based approach, such as CWA early within the design process can act to focus subsequent design considerations. The value of completing the CWA early within the design cycle became evident within Chapters 4 and 5, where it was used to constrain ideas generated with both DwI workshops. In order for an idea to be taken forward for the generation of interface ideas, it had to be mapped to the constraints highlighted with the CWA. This acted to ensure that such an interface was plausible, considered the environment the end-user would find themselves in, and would have a positive impact on emission reduction.

To design an effective eco-driving assistance system, it is important that the system has some measure of adaptability and can be tailored to the driver's particular driving style, as drivers vary significantly in acceleration, following, and cornering behaviour. The development of these metrics was the focus of Chapters 3 and 6. This both aids user acceptance and leads to more useful recommendations. Even if it is very difficult to reproduce a driver's style with an artificial model in general, for energy efficiency and user acceptance purposes a driving style can be summarised with a limited set of parameters, which are:

For acceleration and braking:

a_{max} maximum longitudinal acceleration Typical value: 4 m/s^2
b_{max} braking deceleration Typical value: 5 m/s^2

For vehicle-following:

s_{min} minimum car following distance Typical value: 2 m
T_{min} minimum time headway Typical value: 1.4 s*

For cornering:

Γ_{max} maximum lateral acceleration Typical value: 6 m/s^2
Δ driver curvature safety margin Typical value: 4 rad/km^{-1}

*We note here that this value is less than the "two-second rule" often advocated for safety in driving/highway codes – this "typical" value is representative of what is observed and not the ideal.

Designing for Eco-Driving

These parameters were chosen from existing models of driver behaviour and may readily be estimated from naturalistic driving data or sensors on-board a vehicle. For more information on estimation procedures, we direct the reader to Sections 5.1 and 5.2 of Chapter 3 and Section 4 of Chapter 6.

The importance of these measures to eco-driving optimisations cannot be understated. Truly "fuel-optimal" or "energy-optimal" acceleration and braking in a vehicle is often unnatural for human drivers, requiring long coasting down periods before a stop for example, and is difficult to reliably achieve in practice due to traffic conditions and an ever-changing road environment. Instead, we have advocated blending the objective of energy efficiency with the preferences of the driver. One way to achieve this is by using the optimal control framework for driver preferences introduced in Chapter 6, which formed the basis of the prototype eco-driving assistance system developed during the research project.

The design of an appropriate interface to display information originating from this assistance system to the driver was the focus of Chapter 4 and Chapter 5. These chapters saw the introduction of the Design with Intent (DwI) method for the ideation process. DwI offers an innovative semi-structured approach to support ideation, which seeks to encourage users to be free-thinking to generate novel ideas. This creativity could however result in ideas that are not compatible with the design constraints present within the road environment as outlined by CWA. To compensate for this, it was proposed that the use of mock-ups and potential driving scenarios are key in order to encourage early, but appropriate interfaces to be developed. The ability to gather early feedback from potential end-users allows ideas to be incorporated into future design and development decisions. It was identified that the combined use of CWA and DwI in the ideation process is reliant on a divergence and convergence paradigm. This allows for the creation of multiple ideas which can subsequently be deselected and combined to generate potential interfaces. This down-selection process is essential to allow for focus on useful potential interventions which are both supported by the constraints as outlined within CWA and also possible from a technical perspective. It was noted that DwI can offer a variety of insight and potential useful design suggestions beyond those immediately of value for use in testable interfaces. These ideas can make DwI a powerful tool to use early in a development cycle and can lead to useful spin-off ideas. DwI is a useful design tool, particularly for those working within larger multidisciplinary teams and facing large-scale design problem spaces. This use is based fundamentally on the 138 ideas generated from the DwI toolkit in a single session. Ideas that were generated related either directly to sensory modalities, that is visual, auditory, and haptic, and could therefore be taken forward for interface design, but also related to other notions that designers may wish to consider, potentially acting as kernels of long-term development ideas. DwI, therefore, has both immediate and potential long-term value in developing ideas.

Chapter 7 described the workings of the developed assistance system, which solves a finite-horizon optimal control problem in real-time, in the context of an electric vehicle. Receding horizon control (i.e., nonlinear model predictive control – MPC) can be used for practical implementation on the vehicle. However, there is a mismatch between the problems of minimising energy consumption over an entire journey versus doing so only over the next 30–60 seconds. This is caused by the lack of a suitable "boundary condition" – in the full-journey problem, we may specify that we must get to our destination, but in the receding-horizon formulation this is impossible. The remedy to this is to introduce a terminal cost that penalises reductions in average speed. This has an intuitive explanation: if we go slower *now* then we must speed up *later* to get to the destination on time. Speeding up on later parts of the journey implies greater future energy losses (e.g., to air resistance), so we must account for these.

Once a suitable assistance system had been developed, it needed to be empirically tested within controlled laboratory conditions to ensure its effectiveness and to gain early user feedback. This was the focus of Chapter 8, whereby a driving simulator was used to test the effectiveness of the assistance system with a body of potential users. The design and objects of these initial studies must be carefully considered, and sufficient data must be collected from these trials to inform future development. From this chapter, it is clear that despite their limitations, driving simulators can be used to test potential interfaces and underlying assistance technology as part of the wider evaluation cycle. It was seen from the initial results that although individuals can achieve eco-driving with limited guidance, this was viewed negatively and associated with an increase in driver workload. Initial testing of the assistance system suggests that its use can not only support a reduction in fuel/energy usage but can do so without the negative workload connotations associated with unsupported eco-driving. Simulation trials have a key advantage in that consideration of the results of these trials allow for future refinement of the assistance system to enable it to be more efficient and effective at reducing fuel use and subsequent emissions.

Much current eco-driving advice considers "generic" eco-driving which primarily concentrates on lowering vehicle energy losses, for example due to air resistance and heavy braking, and ignores the effect of the powertrain itself. In Chapter 9 we put our assistance system to the test in simulation for different powertrains for which it was not specifically designed. We discovered that greater energy-efficiency improvements were possible for electric vehicle powertrains, implying that the best results for electric vehicles can only be achieved by tailoring the assistance system to the specific powertrain. With increasing sales of electric vehicles across the world and recent commitments from governments and manufacturers to phase out combustion-engine driven vehicles, this is an important finding for future eco-driving efforts.

Finally, Chapter 10 explored the practical challenges in implementing such an assistance system on-road. In case a detailed powertrain model is unavailable, it is possible to instead develop a predictive model of fuel consumption for the assistance system using a linear regression considering vehicle speed and acceleration. The assistance system is effective in reducing energy demand, and can do so without lowering average speed, but system performance is limited in practice by the availability of reliable sensor data on inter-vehicle distance. With current developments in sensing from autonomous vehicle research, we believe that these difficulties will soon be surmountable.

FUTURE WORK

In terms of the system implementation, future work should concentrate on improving availability of position and speed measurements for the lead vehicle by the incorporation of alternative sensors. The computation time required for processing

Designing for Eco-Driving

and solution of the receding-horizon control problem could also be improved, for example through the use of code generation, leading to an increased update rate and smoother movements of the band on the visual interface. This would open up other possible applications such as direct control of vehicle speed in an eco-driving adaptive cruise control system, and into future autonomous vehicles.

From a research perspective, a detailed study of how the system affects driving style would give insight into the mechanisms involved in fuel-saving and could lead to improvements in the system via encouragement of specific energy-efficient behaviours, such as coasting. This is a complex topic that requires further collaboration between engineers and human factors researchers, and it is clear that the most benefit and route to larger impact is when disciplinary insights are combined. Whilst educating drivers and developing novel technology can both have an individual impact on emissions, it is only by a combination of man and machine working to achieve a common goal that wide-scale change can be achieved.

SUMMARY OF GUIDELINES, BY CHAPTER

The following list of guidelines presents a series of key lessons learnt when tackling the problem of everyday fuel efficiency. It is hoped that consideration of these guidelines can act to support developers, engineers, and researchers, both human factors and automotive, working within this domain.

CHAPTER 1

- Changing driver behaviour can have a significant impact on reducing the fuel used and subsequent emissions when driving.
- Literature demonstrates that numerous approaches can be used to reduce fuel usage when driving, most prominent of these are developing and adopting more technologically efficient drivetrains, eco-driving training, and the use of in-vehicle interfaces to encourage greater engagement with eco-driving techniques.
- Behaviours typically associated with eco-driving include gentle acceleration, minimisation of braking and, early gear changing, to drive at low RPM.
- When considering in-vehicle interfaces, auditory feedback is considered to be disliked and ineffective, whereas both visual and haptic feedback are more readily accepted, despite safety concerns.

CHAPTER 2

- Cognitive Work Analysis (CWA) is effective for generating design constraints for eco-driving systems.
- The specific needs of a driver at any given moment are a greater constraint to potential actions than the type of road the vehicle is travelling.
- The task of driving is prescriptive, and fuel savings are rarely achieved by varying the mechanical actions which a driver can take, savings rather

come from changing the timings of actions and the force by which actions are completed.
- Tasks currently completed by the driver may be more fuel efficiently reallocated in the future to other agents, including automated vehicle functions.
- The use of a structured and evidence-based approach early within the design process can act to focus on subsequent design.

CHAPTER 3 AND CHAPTER 6

- Evidence of real-world driving style should be used as a basis for all interventions targeting changing driving behaviours.
- Evidence shows there is a lot of variation between drivers in terms of their driving style, so for user acceptance, it is necessary to adapt the system to the specific driver.
- Quantitative metrics that help to identify/categorise driving styles are:
  - For car-following behaviour

a	maximum longitudinal acceleration
b	braking deceleration
s_0	minimum car-following distance
T	minimum time headway

  - For cornering behaviour

Γ_{max}	maximum lateral acceleration
Δ	curvature safety margin

- As they correspond to physical properties in most cases, it is typically straightforward to estimate these parameters in the driver model using the same techniques used to estimate the corresponding parameters in the Intelligent Driver Model and Reymond's model (see Chapters 3 and 6 for details).
- Absolute minimisation of the fuel or energy usage requires acceleration and braking profiles that are unnatural for human drivers, for example those that have long periods of coasting down.
- Nonetheless, by blending energy efficiency requirements with typical driver preferences it is still possible to keep energy usage low, without requiring unnatural behaviour. This is possible by exploiting an optimal control framework.

CHAPTER 4 AND CHAPTER 5

- DwI can offer an innovative semi-structured approach to support ideation.
- Mock-ups and potential usage scenarios allow for early feedback from potential end-users, which can be incorporated into future design and development decisions.
- DwI can offer a variety of insight and potentially useful design suggestions beyond those of value for immediate deployment. This can make it

Designing for Eco-Driving 233

a powerful tool to use early in a development cycle and can lead to useful spin-off ideas worthy of pursuit in their own right.
- The use of DwI in the ideation process is reliant on a divergence and convergence paradigm, allowing for the creation of multiple ideas, and then deselecting and combining ideas from the process.
- The down-selection of ideas is essential to allow for focus on useful potential interventions.

CHAPTER 7

- To implement the optimal control framework of Chapter 6 in practice, it is necessary to use some computationally tractable approximation, such as receding-horizon control (i.e. – MPC).
- Performing the optimisation considering the entire journey can further improve energy efficiency; this is impossible in practice, but a similar result may be obtained by applying a terminal cost within the receding-horizon control.

CHAPTER 8

- Driving simulators can be used to test potential interfaces as part of the evaluation cycle.
- Effective eco-driving can be achieved with limited guidance from a predictive assistance system.
- The adoption of eco-driving techniques is necessary to reduce fuel use and emissions but not sufficient for long-term behaviour change and needs to be supported.

CHAPTER 9

- Energy efficiency can be pursued at two levels: reducing "vehicle" losses ("generic" eco-driving) and reducing "powertrain losses".
- Generic eco-driving reduces losses in all vehicles, regardless of the type of powertrain.
- However, achieving the best energy economy requires consideration of the specific powertrain used in the vehicle (see also Chapter 7)

CHAPTER 10

- Predictive eco-driving assistance is effective in reducing energy demand.
- Eco-driving assistance can reduce energy demand without reducing average speed. If we may also reduce the speed, the energy demand decreases further.
- Feeding accurate sensor data to the eco-driving assistance system is key to achieving good results.

Author Index

A

Abrahamse, W., 7, 51, 72, 88, 95, 111
Achour, H., 143
Adell, E., 11, 12
Ahlstrom, U., 21, 33
Alam, M. S., 1, 3
Alexander, C., 96
Allison, C., 14, 20, 40, 71, 78, 82, 88, 89, 95, 110, 144, 172, 173, 187, 194, 198, 214
Allport, F. H., 10
Andersson, J., 133
Ando, R., 2
Andrieu, C., 6, 72
Asadi, B., 206
Attari, S. Z., 72, 95
Azzi, S., 8, 9, 11, 117

B

Bando, M., 121
Banfield, R., 74
Banks, V. A., 188
Barkenbus, J. N., 2, 3, 4, 5, 14, 19, 20, 27, 29, 51, 71, 93, 94, 109, 110, 117, 169, 170, 193, 201, 205
Barker, R. G., 72, 95
Barth, M., 3, 71, 205, 223
Baxter, J., 53
Bellman, R., 119, 120
Benmimoun, M., 50
Berntsen, T., 2
Beusen, B., 1, 3, 6, 110, 117, 170, 193
Biassoni, F., 50
Bin, S., 2, 19, 94
Birol, F., 4, 170
Birrell, S. A., 9, 11, 12, 21, 22, 94, 117, 170
Bock, H. G., 120
Boer, E. R., 53, 64, 66
Boriboonsomsin, K., 3, 7, 71, 170, 205, 223
Bosetti, P., 54, 61, 62, 63, 122, 123
Botkin, D. B., 169
Brackstone, M., 52, 53, 63
Brown, I. D., 50
Brown, T., 111
Buchanan, R., 72, 96, 111
Buckeridge, D. L., 71, 88, 94, 98
Burnham, G., 121, 123

C

Cacciabue, P. C., 20
Camacho, E. F., 157
Carsten, O., 20, 50
Cash, P., 93
Castaings, A., 143
Chan, C. C., 3, 51, 71, 94, 169
Chandler, R. E., 51, 52, 120
Chapman, L., 71, 88, 110
Chen, F., 50, 143
Chen, Z., 144
Cheng, Q., 205
Cornelissen, M., 21
Crawford, J., 9
Crumlish, C., 96

D

Daalhuizen, J., 93, 110
Dahlinger, A., 110
Daun, T. J., 205
Davidsson, S., 9, 32, 33
De Bono, E., 96
DeFazio, K., 54
Degraeuwe, B., 6
Dehkordi, S. G., 170
Delhomme, P., 4, 170, 205
Department for Environment Food
    and Rural Affairs, 71
Department for Transport, 19, 36, 76
Dib, W., 194
Dietz, T., 1, 19
Ding, F., 118, 206, 223
Dogan, E., 8, 9, 170
Dula, C. S., 5

E

Eckert, C., 96
Ehsani, M., 26, 27
Evans, L., 117

F

Fajri, P., 149, 150
Fénix, J., 20
Fidel, R., 21, 74
Fischer, G., 169
Fleming, E., 37

Fleming, J., 50, 118, 119, 120, 144, 146, 147, 170, 171, 174, 194, 198, 202, 206, 207, 212, 213, 214
Franke, T., 5, 94, 193, 202
Froehlich, J., 7, 8, 49, 51, 170
Fuglestvedt, J. S., 1, 2

G

Gardner, G. T., 94
Gibson, J. J., 72, 95, 96
Gillespie, T. D., 27, 28, 29, 54, 187
Gipps, P. G., 121
Godthelp, H., 53, 122
Godthelp, J., 53
Gregg, J. S., 2
Guzzella, L., 176

H

Han, J., 144
Hargraves, C. R., 120, 133
Hart, S. G., 175, 176
Harvey, C., 103
Hatcher, G., 72
Heijne, V., 51, 117
Herrey, E. M., 51
Hess, R., 53
Heyes, D., 3, 6, 7, 170
Heyman, J., 8, 9
Hill, N., 1, 2, 19, 71, 94, 98
Hooker, J. N., 4
Horberry, T., 176
Horrein, L., 27
Houghton, R. J., 35
Houska, B., 214
Hu, J., 194
Hu, S., 143
Hülsheger, U. R., 6
Hung, Y. H., 143
Husnjak, S., 10

I

IEA., 169
IJsselsteijn, W., 95

J

Jain, R. K., 72, 95, 111
Jamson, A. H., 12, 29, 72, 95, 111, 188, 205
Jamson, S. L., 176, 187
Jenkins, D. P., 21, 22
Jin, Q., 118, 206, 223

K

Kozak, K., 49, 66
Kalman, R. E., 120
Kalyuga, S., 37
Kamal, M. A. S., 118, 144, 193, 202, 206
Kant, V., 21
Karl, T. R., 71, 88, 110
Ke, W., 143
Kesting, A., 52, 121, 131
Kiefer, R. J., 63, 64
Kilgore, R., 21, 37
Kircher, K., 9, 10
Kirk, D. E., 45
Knowles, M., 193, 201
Krauβ, S., 121
Kurani, K., 1, 2, 3, 4, 71, 170
Kuriyama, M., 148

L

LaClair, T. J., 27, 28, 29
Lai, W. T., 6
Larsson, H., 11
Larue, G. S., 170
Lauper, E., 7, 51
Lawson, B. R., 111
LeBlanc, D., 49, 65, 66
Lee, D. N., 53, 63, 64
Lee, H., 187
Lee, J. D., 50, 66
Leutzbach, W., 53
Levison, W. H., 54, 61, 62
Li, L., 71
Li, S. E., 117, 164
Lin, X., 205
Lindgren, A., 50
Lockton, D., 71, 72, 74, 77, 78, 88, 93, 95, 96, 98, 109, 110
Lorf, C., 3, 94, 169
Lot, R., 131

M

Ma, H., 117
Madden, T. J., 8
Madhusudhanan, A. K., 206
Marchau, V., 50
Martin, E., 2, 71, 170
Mayne, D. Q., 159
McCubbin, D. R., 71, 88
McGrenere, J., 72, 96
McIlroy, R. C., 20, 21, 22, 33, 37, 39, 40, 71, 75, 94, 95, 103, 110, 111, 169, 170, 193

Author Index

McMichael, A.J., 169
McRuer, D. T., 53
Mensing, F., 3, 7, 72, 95, 118, 170
Meschtscherjakov, A., 1, 7, 8, 51, 170, 173, 193, 202, 206, 207
Miller, S. A., 28
Monteil, J.,118

N

Naikar, N., 21, 31, 35, 40, 74, 75
Netherlands Environmental Assessment Agency, 2
Neumann, I., 193, 202
Nielsen, J., 89
Norman, D. A., 72, 96

O

Oinas-Kukkonen, H., 20
Ojeda, L. L., 144
Orfila, O., 205

P

Paganelli, G., 196
Pampel, S. M., 20, 170, 187, 231
Pannells, T. C., 93
Parasuraman, R., 49, 66, 118
Philipsen, R., 94
Piatkowski, D., 169
Pick, A., 53
Pontryagin, L. S., 119

R

Rakha, H., 117
Ramanathan, V., 93, 94
Rasmussen, J., 9, 20, 24, 31, 37, 74, 88, 100, 111
Read, G. J., 74, 88, 89, 98
Recarte, M. A., 9
Redström, J., 72, 95
Revell, K. M. A., 95
Reymond, G., 50, 53, 54, 60, 62, 63, 64, 65, 66, 118, 122, 127, 131, 135, 147
Rittel, H. W., 111
Robinette, D., 27
Rolim, C., 10
Rose, G., 169
Rouzikhah, H., 9, 10

S

Saboohi, Y., 4, 118, 170
Salmon, P. M., 96, 98, 103
Sato, Y., 147
Schall, D. L., 144

Schunn, C. D., 99
Schweitzer, F., 72
Sciarretta, A., 117, 119, 176, 206
Seminara, J. L., 49
Shabbir, W., 5, 143
Sharp, T. D., 22
Simon, H. A., 53, 72, 95
Simonelli, F., 206
Sivak, M., 3, 5, 14, 94
Skeie, R. B., 2, 169
Sorkin, R. D., 49, 50, 66
Stanton, N. A., 4, 14, 20, 21, 22, 31, 33, 37, 39, 40, 71, 74, 75, 77, 78, 82, 88, 89, 95, 103, 110, 173, 176
Staubach, M., 13, 49, 51
Stillwater, T., 2, 71
Strömberg, H., 2, 4, 19
Summala, H., 9
Sun, C., 144

T

Takezaki, J., 29
Tate, E., 20, 193
Thatcher, A., 93
Thornton, J., 1, 19, 93, 98
Tidwell, J., 96
Todosiev, E. P., 53, 56, 63
Treiber, M., 52, 63, 118, 121, 131, 146, 147
Trommer, S., 206
Tromp, N., 93, 110, 111
Tulusan, J., 7, 8, 20, 72, 95, 110, 170
Tunnell, J., 28, 29
Turrentine, T. S., 170

V

Vaezipour, A., 170
Van der Voort, M., 8
Van Erp, J. B. F., 11
van Westrenen, F., 21
Van Winsum, W., 49, 52, 63, 64, 122
Vandenbergh, M. P., 1, 2, 19, 94, 117, 205
Vatanparvar, K., 149
Vicente, K. J., 21, 22, 24, 33, 35, 37, 39, 75, 100, 111
Vining, C. B., 51, 169
Viviani, P., 54

W

Wachter, A., 133
Wåhlberg, A. E., 1, 6, 14, 143
Wan, N., 144
Wang, F., 143
Wang, J., 50, 63, 64, 65

Wang, M., 117
Wei, Z., 143
Wever, R., 93
Wickens, C. D., 11
Wilson, R. E., 121
Windecker, A., 3, 169
Wu, C., 143, 193
Wu, Y., 20

X

Xiang, X., 118
Xu, Y., 144

Y

Yan, X., 29, 55, 131, 147, 148, 151, 163, 194
Yang, Q., 52
Yilmaz, M., 169
Young, K., 9
Young, M. S., 9, 20, 94, 170

Z

Zarkadoula, M., 6, 14, 117
Zhao, L., 144
Zhao, X., 12, 13

Subject Index

A

Acceleration, 5, 11, 34, 52, 121, 146, 180–182
  lateral acceleration, 54, 60, 122, 127, 147
Advanced driver assistance system (ADAS), 49–51, 66, 118
Alternative transportation schemes, 169
Autonomous vehicles, 64, 193, 223, 231

B

Braking, 5, 33, 35, 148–150, 182, 183
  regenerative braking, 148–150

C

Cognitive Work Analysis (CWA), 20–22, 74–77, 88–89, 228–229
  Contextual Activity Template (CAT), 21, 31, 32, 35–37, 41–43
  Control Task Analysis (ConTA), 21, 30, 33, 75
  Skills, Rules and Knowledge taxonomy (SRK), 21, 37–39, 42, 44
  Social Organisation and Cooperation Analysis (SOCA), 21, 35–37, 42–43, 75
  Strategies Analysis (StrA), 21, 33, 40, 42–43, 75
  Work Domain Analysis (WDA), 21–22, 30, 39–41, 43, 75, 82
  Worker Competencies Analysis (WCA), 21, 37, 40, 42–44, 75
Collision warning, 49–50
Cornering, 53–55, 59–62, 155–156, 220–221
  cornering speed, 53–55, 59–62, 66, 122, 127–128

D

Design guidelines, 231–233
Design with Intent (DwI), 72–79, 81–82, 84–89, 95–99, 110, 228–229
  lenses, 72–73, 79, 84, 96, 97, 100, 110, 112
Design workshop, 76–78, 98–99
Desired velocity (speed), 123, 186, 223
Distance to leader (headway), 51–53, 130, 146–147, 153–154, 163–164
Driver acceptance, 49–50, 66, 103–105, 228

Driver model(ling), 51–55, 120–122
  Intelligent driver model (IDM), 52, 118, 121, 129–130
  Driver satisfaction model (DSM), 123–128, 158
  driver preferences, 117–119, 127, 145–146, 194, 206–207
  validation, 55–62, 131–135
Driving behaviours, 1–9, 20, 64, 94–95, 227
Driving simulator, 171–175, 197–198
Driving style, 2, 94, 202, 228, 232

E

Eco-driving, 2–14, 20, 117–120, 143–145
  assisted eco-driving, 50–51, 143–145, 193–199, 205–207
  environmental impact, 1–3, 71, 93–95
  financial impact, 1, 3–4, 7–8, 14, 82, 170
  interfaces, 13, 20–22, 35, 37, 40, 44, 71–90, 93–112, 117, 169–173
  interventions, 35, 37, 44, 71, 95, 103–109
  knowledge, 3–6
  training, 1, 2, 4, 6–7, 14, 20, 110, 117, 202
  unassisted eco-driving, 2–14, 20, 117–120, 143–145
Electric vehicles, 143, 147–152
Emissions, 1–3, 19–20
  carbon dioxide, 19, 93–94, 117
  nitrous oxides, 19, 93–94

F

Feedback, 7–14, 40, 170, 193
  auditory, 8, 12–14, 40–43, 77, 84–85, 99, 104, 117, 193, 205, 227, 229, 231
  haptic, 8, 11–14, 40–42, 77, 99, 101–102, 104, 117, 188, 193, 205, 223, 227–229, 231
  post journey, 10, 31
  real-time, 10, 50, 52, 73, 81, 97, 118, 133, 172–173, 193 198, 205–206, 215, 224
  visual, 8–14, 40–43, 82, 86, 99–108, 169–189, 202, 205, 207, 209, 214–215, 223–224, 227–231
Following distance, 51–53, 130, 146–147, 153–154, 163–164
Fuel/Energy consumption, 5–7, 117, 143–144, 147–150

G

Greenhouse gas (GHG) emissions, 1–2, 19, 93–94, 117

H

Hybrid vehicles, 143, 195–197
Hardware, 28–30, 207–209
    ADAM/data logger, 55
    CAN bus, 209
    GPS, 28, 30, 208–209
    radar, 29, 209

I

Instrumented vehicle, 218–222

N

NASA-Task Load Index (TLX), 175–176, 184–185
Naturalistic driving, 50
    car following (vehicle following), 51–53
    cornering behaviour, 53–55

O

On-road testing, 218–224
Optimal control, 119–120, 128, 135–137
    full-horizon optimisation, 152–153
    receding horizon control/MPC, 157–162, 213–214

R

Road geometry, 60
    road curvature, 54, 122, 127–128
    road slope, 148, 209

S

Safety, 9, 50–51, 122, 223
Speed advisory, 144–146, 152, 158
Speeding, 11, 224

T

Time headway (to leader), 51–53, 130, 146–147, 153–154, 163–164
Time-to-collision, 57–59, 63–64, 122
Time-to-lane-crossing (TLC), 53, 122
Traffic, 3, 30–32, 51–52, 63, 144, 172
Transport pollution, 19, 71

V

Visual HMI, 171–174, 214–215
Vehicle model, 146–152
    energy consumption model, 147–150, 197
    energy losses, 148–152
    fuel consumption model, 196, 211–213
    powertrain, 147–152, 193–197
    regenerative braking, 148–150, 193, 197